D0463370

WITHDRAWN

CATTLE

An Informal Social History

CATTLE

An Informal Social History

Laurie Winn Carlson

GARRETT COMMUNITY
COLLEGE LIBRARY

Ivan R. Dee

CHICAGO 2001

CATTLE: AN INFORMAL SOCIAL HISTORY. Copyright © 2001 by Laurie Winn Carlson. All rights reserved, including the right to reproduce this book or portions thereof in any form. For information, address: Ivan R. Dee, Publisher, 1332 North Halsted Street, Chicago 60622. Manufactured in the United States of America and printed on acid-free paper.

The excerpt from *With These Hands* by Joan M. Jensen is used with the kind permission of the Feminist Press.

Library of Congress Cataloging-in-Publication Data:
Carlson, Laurie M., 1952–
Cattle : a social history / Laurie Winn Carlson.
p. cm.
Includes bibliographical references (p.).
ISBN 1-56663-388-5 (alk. paper)
1. Cattle—History. 2. Cattle—Social aspects. I. Title.
SF195 .C37 2001
636.2'009—dc21 2001023998

For my parents, Ed and Juanita Winn

Contents

Preface

WHEN I BEGAN WRITING THIS BOOK I had no idea that a topic as general as cattle could be so polarizing. Yet it has been. I found that if I was with a group of environmentalists and mentioned I was at work on a book about cattle, they turned cool and distrustful, and distanced themselves from me. In a group of Western ranchers I received the same response. When I talked about eating meat, my vegetarian friends were as ideological about their eating practices as my beef-eating acquaintances. I began to sense an awkwardness about being in the middle; few people are on the fence, so to speak, about the topic of cattle grazing, eating beef, or drinking milk. For me this provided an additional challenge: it became more difficult to write in a way that would please everyone. I realized that my attempts to synthesize and find meaning from both camps might simply alienate all readers. I hope this is not the case. I hope that wherever readers may stand ideologically on the cattle issue, they will find something of value in this book.

Covering thousands of years of history in one small volume has at times seemed overwhelming to me. Critics will no doubt find gaps in my survey, and nuances that have been overlooked. My goal is to arouse readers' curiosity by dangling some interesting facts before them, to urge them to see history a bit differently than they might have before, and to see how cattle have influenced our human history.

This book took root one autumn a few years ago when my husband and I attended the Spokane County Fair. We had both been raised on small farms where we had grown up with a few cows, and we had both drunk milk from the family cow. But as adults we had not thought much about cattle. We had left all that behind us when we went off to college and careers. Not until years later, when we attended the fair specifically to see a friend's hybridized gladioli entry, did we encounter cattle again. We wandered through the midway and commercial areas—loud, raucous, and overhyped—until we noticed a small stream of people heading into one of the exhibit buildings. We followed to see what was going on. An auction was about to begin for 4-H youths' animal projects: hogs and steers would be sold. Youngsters and their families were milling about in the stands, finding good seats or giving their animal a final grooming touch.

We headed out a door on the side of the building and into the adjoining barn where I was immediately struck by what can only be described as an epiphany. The big open-ceilinged barn was filled with cattle—beautiful, peaceful, noble cattle. It was a hot September day, mid-afternoon, so they were all lying on fresh straw bedding, chewing with eyes half-closed, nodding away in naps. The animals we first came upon took my breath away: they were absolutely beautiful and looked so unusual and eerily historic that I realized I had never seen anything like them. Smaller than the 4-H kids' grain-fattened steers, they were nicely compacted yet still big animals. They had chestnut-colored hair, which hung shaggily and framed beautiful faces with wonderfully dreamy dark bovine eyes. They were Scottish Highland cattle, and walking among them was like taking a trip back to some place in antiquity, to some ancient heritage within me that I had never tapped. A young boy, about ten years old, was lying alongside one of the animals, both of them sprawled peacefully on the soft straw, his arm draped over the sleeping animal's back. A tape player atop some hay bales was playing Celtic music.

The animals were utterly beautiful, something I had never realized about cattle before. Over the years I had been delighted with Gary Larson's cow cartoons, had scoffed at cows for their slowness and stupidity, and had thought them simply dumb or irrelevant, perhaps dispensable. After all, cows were, well, *cows.* I had never realized that the animal itself could be so enchanting.

We stayed awhile, watching the animals, their tenders, and other fairgoers who admired the cows as much as we did. Then we moved on to leave the building—and were surprised by a reminder of the here and now. At every exit from the cattle barns, the Spokane County Health Department had installed hand-washing stations. Signs warned everyone to wash their hands with the antiseptic soap that was provided, in order to prevent life-threatening infection from what had become a national threat, *e. coli.* Certainly it was not overkill: that very week several fairgoers in an Eastern state had sickened with *e. coli* after drinking water that had been contaminated by drainage from fairgrounds cattle barns.

How, I wondered, did we get to this point, where these beautiful animals that humans had relied on for centuries, these noble and gentle creatures, could now so innocently mean death to us? How had we become so far removed from these ancient animals that we had allowed—even caused—them to become so dangerous to us? We washed our hands thoroughly and moved on, working our way through the crowd outside and the hawkers of greasy fried-dough treats, wireless phone services, and Yamaha motorcycles. But I continue to hold in my mind that peaceful scene: a young boy draped over a napping cow, both trusting and caring for each other, as if in some tranquil meadow far back in antiquity. Our relationship with cattle goes back for millennia. The story of how we have changed each other is fascinating.

The better I understood them,
the more complex and worthy of study
did their minds appear to me.
—Francis Galton, writing about
cattle in Africa, 1871

CATTLE

An Informal Social History

PART ONE

PART OF THE FAMILY

1

Cows on the Ceiling

Not one of us could paint like that.
—Picasso, speaking of
Altamira's images

IT WAS ONE of the most extraordinary finds of the century, one that completely changed the way humanity, biology, and cultural evolution had been understood. And it was due to a child.

In 1879 little Maria Sautuola, aged somewhere between five and nine years, accompanied her father on one of his archaeological forays on the family estate in the Cantabrian Hills along the coast of northern Spain. Marcelino Sanz de Sautuola had been caught up in the fever for archaeological discovery that swept western Europe in the late nineteenth century. He had attended the Paris Exposition of 1878 and was as intrigued as the rest of Europe about the recent finds from the Ice Age that were displayed there. Like many other educated individuals of the day, he cultivated an interest in natural science. Almost all archaeologists of the day were amateurs. Most were wealthy men who possessed the time and enthusiasm to explore the secrets of the natural world.

The Sautuola family landholdings included a large cave, which Marcelino hoped might hold something of potential importance. He began by himself to excavate the area at the front of the cave where he had recently found some stone and bone implements. His daughter Maria accompanied him and entertained herself as most young children would: she wandered off, looking for anything interesting. She later said that as her father excavated the cave floor, she had been "running about in the cavern and playing about here and there. . . . Suddenly I made out forms and figures on the roof." She shouted to her father to come look. "*Mira, Papa, buyes!*" ("Look, Papa, bulls!"). She had immediately identified the figures as images of oxen.[1]

Human eyes had last seen the paintings on the ceiling of the Spanish cave at what became known as Altamira nearly fourteen thousand years earlier. The entrance had fallen in, and the Sautuola family had not even known of the cave until a hunter stumbled onto it and alerted them to its existence.

Altamira cave is no simple hole-in-the-wall as caves go. On a slight incline three miles from the crystal-blue waters of the Bay of Biscay, it is a big, sprawling labyrinth of rooms with paintings scattered throughout the entire thousand-foot length. The chamber where Maria spotted the famous cow ceiling is to the side of the first room, which was the general living area. Now known as the Great Panel, the mural extends across the ceiling and portrays more than thirty large bison and a few other animals. Many of the animals are painted on curving stones so that their shape fits the rounded stone, adding a dimension of depth to the images. The animals are painted as if in a herd, some standing still, some galloping, some bellowing, some even rolling in the dust. Some are more than six feet from nose to tail; one large deer is painted life-size. Other, simpler images are scattered elsewhere—hand-prints, symbols, images of horses, deer, and goats. There are also engravings of deer, goats, horses, and two additional bovines. The room that holds the Great Panel is vast—more than fifty feet across. At the back of the cave are other chambers; the last has all its exits flanked by masks painted on the wall in the shape of cattle heads.

The artists and their audiences left litter all about the cave; remains of their meals include mollusks, deer, goats, and small mammals. Paint supplies were discarded too, including ochre, pebbles used to grind mineral colors, and other tools.

SAUTUOLA REALIZED that the cave and its unusual paintings were significant, but because he was not a professional or scholar his information was ignored, even denigrated by the scientific community in Madrid. In an era when interest in the Ice Age had prompted a trade in fake artifacts in southwest France, academics considered Altamira to be something that Sautuola had created by himself. Some accused him of hiring a mute local artist who would be unable to testify against Sautuola—to fake the works. Afraid it was simply another hoax, no one took him seriously.[2] An expert, Dr. E. Harli, was commissioned to give his opinion. It was not favorable; he confirmed the suspicion that Sautuola may have created a fraud. The paint looked too fresh, too bright, and there were no soot layers. Surely if prehistoric people had painted these things, the soot from eons of campfires would have clouded the surface of the paintings.[3]

Sautuola continued to seek the validation of professionals, begging them to view the cave at Altamira. Scholars belittled and disdained him; only one, a professor of geology and paleontology at the University of Madrid, recognized the importance of the discovery. Juan Vilanova y Piera visited the site, then presented his findings at conferences in Portugal, Germany, France, and Spain. But audiences scoffed and sometimes met his statements with open contempt.[4]

Nothing like Altamira had been found before; it was too big, too old, of such quality—it simply could not be authentic. Scholars cried hoax, blaming Sautuola, who was thought to be in league with local villagers seeking tourists. One scholar charged that the cave paintings were an attempt by the conservative Catholic church to ridicule the emerging sciences of paleontology and prehistory. After all, the beauty of the paintings thwarted the thesis of evolution as a path to perfection; classical scholars could not connect the paintings to the art of Greece, Rome, or Phoenicia.

The cave art had been discovered at a time when evolutionists (which most prehistorians in the 1870s had become) could not link these spectacular findings at Altamira to the science of evolution. They were of such grandness and magnitude, so elaborate and beautiful, that they could not possibly be the work of Ice Age man—a being that scholars at the time held to be little more than an animal that walked upright. The ancient paintings had to be reconciled with the idea of humans living more than fifteen thousand years in the past, a prospect not easily digested in the mid-nineteenth century. Stone and bone tools found in France had been shocking enough, so shocking they were a major draw at the Paris Exposition in 1878, which drew visitors such as Sautuola himself to see the unbelievable finds.

For more than twenty years the discovery at Altamira sank into oblivion. If not for Sautuola's dedication, the murals might have been destroyed. The paintings at Altamira were simply too much, too soon. Scholars could not accept the idea that ancient people possessed the sophistication, refinement, and creativity to create such artistic expression. Only after several other painted caves were discovered in southwest France, notably at Lascaux, was this early art finally accepted as legitimate. Emile Cartailhac, the leading prehistorian of the era, published an apology to Sautuola and admitted that his mistaken ideas had kept him from visiting the cave until twenty years after Sautuola alerted him to its existence. Cartailhac's "Mea Culpa of a Skeptic" was published in 1895, immediately following his long-last visit to Altamira.

Cartailhac came too late for Sautuola, though. The cave's longtime owner and advocate for Altamira's art had died seven years earlier. Before his death he had self-published a book about his discoveries in the cave and had spent the rest of his life disappointed that his findings were ignored. The aging Professor Cartailhac did meet with Maria Sautuola, now a grown woman, who had first seen the "cows on the ceiling" nearly twenty years earlier.

The findings at Altamira and Lascaux, as well as the many other cave art sites in western Europe, caused modern society to rethink their assumptions about prehistoric people. In the surge to embrace Darwinian evolutionary concepts, scholars had regarded these people as much farther from modern humans on the evolutionary scale. Even in the nineteenth century, ethnologists had viewed "primitive" peoples as uncivilized, an opinion rooted in European superiority. If evolution meant that a species grew more refined and perfect, how could these beautifully executed pieces of art fit a theory based on a continual progression toward perfection?

It was little Maria Sautuola's discovery of the "cows on the ceiling" that changed how we thought about prehistoric people. It made us realize they were far from the animal-like creatures that Darwinian theory seemed to describe. Altamira became the foundation for a transformed field of archaeology—it changed from a field obsessed with collecting material goods, particularly jewelry, to one that examined the cultures of the people who had made the articles.[5]

BECAUSE THE PAINTINGS are on the ceiling of the cave, the slow dripping of moisture that covered them had sealed them like a thick, solid varnish. The painting techniques used at Altamira, and at Lascaux, the major cave site in France, are similar. Experiments trying to copy the cave art have shown that the artist or artists must have used a technique that involved multiple steps. First, a carefully drawn outline of the animal was made, then color pigment was applied by spattering paint against the rock, filling the interior of the outline with various colors and gradations of tint. Most likely the artist used a brush to spatter the paint, the same way children use toothbrushes filled with paint to spatter outlines of autumn leaves. Experiments show that the artist was able to spray a fine layer of paint by spitting the paint from a distance of a few inches away from the rock surface. Michel Lorblanchet, a leading authority on the experimental reconstruction of cave paintings, has found it possible to re-create images such as those at Lascaux and Altamira by spitting ochre and charcoal while using his hands to shield surrounding areas from the paint. This technique produces edges that are either sharp or fuzzy, as desired. It took him about thirty-two hours to do a mural, allowing for drying time between four sessions of paint application. A forerunner of today's airbrush, the technique also may have involved spraying pigment through a hollow tube, perhaps bone, to achieve the finely misted paint application.[6]

"Spit painting" sounds distasteful today, but imagine watching someone applying brilliant pigment in that manner, in a dark cave lit only by firelight. It might appear as if the artist were magically breathing life into the rock as the images emerged from the wall. The artwork had to have a purpose, because the logistics of getting into these caves, which were always dark, demanded that a supply of adequate lighting be brought in along with the pigment and application materials. Carrying these materials into the caves, setting them up, and providing adequate lighting suggests there was a reason to go to all the fuss.

The paints themselves were deliberately and carefully created. Gypsum was the commonly used binding agent for the pigments; talc was used as an extender to make the paint go further. It also improved adhesion to the damp cave rock and stopped the finished painting from cracking. The charcoal used for outlining the images did not come from the campfires near at hand; rather, the artists sought to obtain a charcoal from resinous pine trees that grew at a distance. Pebbles were used to grind pigment, usually ochre and iron oxides. Small mollusk shells that held pigment have been found in the caves, tossed where the artist discarded them. Similar artistic techniques have been noted at Lascaux, where the paint appears to have been blown from a tube, sprayed in a fine mist over a large area. Color-stained bone tubes have been found, still containing remnants of powder pigment. Stone palettes have been found in some caves, still holding remains of red, black, and yellow pigment.[7]

The work was lit by lamps carried into the dark recesses of the caves, made from shallow concave stones or clay molded to fill with animal fat or marrow, and a plant fiber wick. Researchers have re-created these primitive lamps and have kept them lit for more than two hours inside a cave.

The animal images were first meticulously engraved with flint tools. The horns, ears, muzzles, and beards, even the tongues of some are detailed. The uneven rock surface demanded that the artist be patient, controlled, and careful. Once the outlines were engraved, paints were applied to the damp limestone ceiling in a technique similar to fresco painting. There could be no erasing, no crossing out if errors were made—and there are none. The artist's strokes are firm and unwavering; there are no corrections.[8]

Who were the artists? Most were right-handed, that's about all we know definitely about them. The way the paintings and engravings show up when lit from the side, not straight on, confirms that the artists were working with a side light. Many caves also contain images of hands; the artists simply used their own left hand as a stencil, holding it against the cave wall while they spattered paint over it with the right. We do know that the work at Altamira, the beautiful ceiling, was the work of an individual artist, because a personal style shows up in all the images there.[9]

FOR MOST OF the twentieth century, a Catholic priest was the major authority on the cave paintings, particularly those at Altamira. Father Henri Breuil dominated the field of prehistoric art until his death in 1961. He believed the images were based on hunting and fertility magic. For most of the twentieth century this explanation held, until scholars in the 1960s and 1970s began to see references to male-female relationships in the images. Now the art was thought to be made up of symbolic animals—the horse, considered to be a female symbol, and cattle, a male symbol. Other symbols scattered along the cave walls were seen as phallic or sexual representations. This interpretation was itself overridden in the 1980s by a rejection of any all-encompassing hypothesis to explain early human rock art.[10] Today, while we admire the early cave art of western Europe, its images of cattle, horses, and other animals drawn with refined artistic technique, we can only hazard a guess as to why they were done or what they meant.

The art could have been intended as literal, political, religious, sexual, whimsical—even mundane. We cannot explain what may have been an image or group that had several purposes at once. As

Paul Bahn, an authority on prehistoric art, notes, "Even a single motif may have pluralistic meaning. Or it may not. There are pitfalls in an excessively literal interpretation of ancient art, but there are far more in nonliteral interpretations, which, at their worst, are mere wishful thinking or flights of fancy." Bahn favors accepting the images for what they are. "Where ethnographic evidence is absent (as in the majority of cases), literal interpretations of prehistoric art are far safer—and at worst may be incomplete rather than wrong."[11]

We can never know the minds of prehistoric people. Their art may have referred to rites of passage, death, birth, rebirth, tribal secrets, laws, taboos, love, sorcery, spiritual prayers, animal totems, or any shared tribal story. The paintings may have been for entertainment, decoration, or even education. Bahn points out that "our own psychological needs, as well as the reward structure of academia and society, play a considerable role in promoting exotic interpretations of the past."[12]

Distancing ourselves from the exotic, perhaps we should examine the paintings in light of the human relationship to the image most often found in early cave art—cattle. The long-held hunting-magic explanation does not hold up because Eurasian prehistoric art does not always depict animals that were the major food source. People living near the western European cave sites based their diet at that time on small mammals and deer. And in many other parts of the world, rock art shows animals that were not mainstays of the local diet. The animals in the art must have been there for reasons other than hunting.

The paintings show cattle. Not today's cattle but ancient *aurochs,* now-extinct relatives of today's domestic cattle. The aurochs in the paintings are large with very broad heads, short up-curving horns, and a beard. The female is smaller in stature, less bulky, and more gracefully angular. Zoologists note that the images are remarkably accurate. The fact that the images are of *animals,* something scholars have seldom found fascinating, helps explain why it has been so difficult to come to a satisfactory interpretation of Altamira's images. L. G. Freeman, a prehistorian and an expert on the Altamira images, points out that "the most important obstacle to the recognition of the species depicted in Paleolithic art, and the greatest single cause of mistaken identifications and pointless debate in the literature, is the fact that many prehistorians have been (and still are) shockingly ignorant about the appearance and behavior of wild animals."[13] That these animals are the ancestors of domestic livestock makes it even less likely that scholars would find them intriguing.

But Paleolithic people were not so jaded. This art may have provided them a unique way of viewing animals. Their art shows the animals in groups, not as single representations as they might be if they were worshiped as an animal totem.

The artists mixed the species in a composition and showed several of each animal, all in different postures, relating to what they observed in daily life. Scenes of grazing animals were represented by a diversity of figures. Whether or not the artist had other motives as well, the representation of a natural scene of free-ranging animals taken from personal observation provides a basis for interpreting the importance of the images.

In fact, free-ranging animals do associate with one another. Their sociability relates to their species, but even solitary animals come together with others from time to time. The cave artists depicted animals that range together: bovines, horses, goats, deer. The animals live in social groups, just as humans do. Prehistoric cave art, especially in western Europe, never shows animals in predator-prey relationships. Wolves were seldom drawn and bears were never represented, though both species were prolific and humans would have encountered, perhaps feared, them intensely.

Rather, prehistoric artists seem to have been interested in the social dynamics of the animal world. And bovines provide an interesting social dynamic for study. They are herd animals who live together yet willingly share the range with other species. The Great Hall at Altamira shows at least twenty large aurochs—early cattle—all painted by the same artist. They were done at one time and deliberately composed as a group, a herd of sorts. The ceiling composition must be studied as whole, not as individual symbolic representations of an animal. They are all adult aurochs in a social grouping that would occur in nature only during the mating season, when female and young animals are at the center of the herd and adult males roam the fringes for group defense. Occasionally other species appear in the paintings, animals that would have grazed alongside aurochs, such as deer and horses.

At Altamira, the entrance to the room with the Great Panel has three male aurochs facing the door. Females and other animals whose sex is not evident are in the middle of the ceiling, and on the opposite edge of the mural are the images of more male aurochs, their heads facing out the opposite direction. The mural depicts a herd of bovines and their unique herd behavior at mating time. The entire mural is not completely visible from any one spot in the room, so it may have been designed for group viewing—a shared social experience for a group of people.

One figure, an engraving in another part of the cave, has been difficult to explain: the image is clearly of a male bison with another animal directly behind and above. Some have thought it to be a wild beast attacking the bison from behind, others have postulated that it was a primitive attempt at artistic perspective by imposing one animal over another. In the 1970s it was restudied and determined to be an image of bison in the act of mating, but one in which the smaller animal (the female) is mounting the larger. Difficult for scholars to make sense of, it depicts a fairly typical activity among cattle of both sexes. When a cow is ready to mate and the bull is not, it is common for the cow to initiate the activity, perhaps to get the male in the mood if not simply to make her intentions known. Even Aristotle noted this behavior, calling such cows "bull-struck."[14] L. G. Freeman contends that the entire mural at Altamira is a depiction of mating cattle. So too is the art at other locations. Deep in the cave at Tuc d'Audoubert, France, about 750 yards from the entrance, a bas-relief of beautifully modeled clay bison rises from the walls. There, wet clay was modeled directly onto the rock wall to represent two exquisitely detailed bison. Two feet in length, they show a cow with a bull behind and slightly above.[15]

If bovine reproduction is the theme of the huge mural at Altamira, what purpose or meaning can it have? Freeman emphasizes the importance of observing prehistoric art with animal morphology and behavior in mind. The art suggests "that people thought of themselves, as well as the bison, as essentially unthreatened, dominant creatures of their kind in a benevolent and predictable environment."[16] Aurochs and humans evidently stood in close symbolic relationship to each other. Human-bison hybrids are apparent here and there in prehistoric French and Spanish art; it was fairly widespread geographically. Images that combined humans and cattle as one being could have meant that people saw a commonality in the behavior of both species. Perhaps early people adopted similar societal behaviors, with females and males congregating in gender groups rather than as heterosexual couples or in small families. Perhaps patriarchy was preceded by a social grouping similar to that of the herd: males stick together, as do females and children. The two mix when necessary, otherwise they find their social life within a gendered grouping. People living before the rise of clan or tribal social structures may have emulated the most social animals they knew: cattle. Bands of Paleolithic humans may have formed "herds" of their own.

The animals themselves are central to explaining why people found the cave images so fascinating and valuable. The people at Altamira took great pains to create the ceiling murals. Somehow they had not only to create pigments and tools, and develop a style of art, they also had to obtain their models. For the cattle at Altamira and Lascaux were painted using dead aurochs as models. We know this because of the angle of the head and offset pair of legs. But just because they painted from dead animals did not mean they wished to depict animals as dead or as trophies. Instead they sought to show them in dynamic positions.[17] The artists were showing what they valued most about the animals they painted—their behavior. They did not show men throwing spears at the animals, nor images of blood or slaughter. Nor are the animals depicted in a supernatural way, which would have denoted worship.

THE ENVIRONMENT underwent profound changes during the Mesolithic period, 8000 to 2000 B.C., a time of warming temperatures and retreating glaciers. As the ice disappeared, birch and willow forests, later pine, spread northward with the warming climate. Hazel trees sprouted throughout Europe, up to seventeen times more numerous than other trees, with deciduous forests following the pioneering hazels. This spreading undergrowth and forest further diminished grasslands, and herd animals that had fed off them were forced to move elsewhere. Settlements with new ways of hunting had to be developed; smaller bands of people could hunt with the bow (a technology available thanks to the encroaching forests), the dog, and the boat. These three innovations allowed people to survive as the world around them changed and the herd animals shifted to other climes. People formed remote, hostile bands in order to protect or fight over hunting rights to particular geographic areas. It was during this time that the fine art of the cave painters ceased. Both large-scale murals and smaller carved and etched items disappear in forms that have endured until today.[18]

What could be the reason behind the demise of such grand works of public art? Domestication of animals appears to be the catalyst for the shift of art in a different direction. Because it was spread over many centuries, the change was rather gradual. After pottery developed, the artistic medium changed and so did the subject matter. "The domestication of animals went to the bottom of a primeval relationship, altering it at the roots. Man as farmer sees himself and his flocks and herds with different eyes from those with which the hunter watches the wild quarry in the forest. . . . The ploughing ox team and the herded sheep are the creation of the farmer in a sense that the wild goats and cattle which a hunter shoots and traps are not."[19] A shift in human perceptions about cattle occurred along with their domestication. Their mystery faded when they were fastened to the yoke.

Interestingly, these tiny, cringing men and women images that began showing up in art were surprisingly true to life. The adoption of agriculture and its impact on human health produced people who were smaller and less healthy than their ancestors. Agriculture provided a different and more inadequate diet for early crop growers. No longer gorging themselves on rich red meat, people subsisted on mushlike meals made from cooked grains. Without adequate protein, dietary deficiencies from rickets to mental depression set in. Farming was killing them.

2

The Domestication of Cattle

Although the ox has so little affection for, or
individual interest in, his fellows, he cannot
endure even a momentary severance from
his herd. If he be separated from it by
stratagem or force, he exhibits every sign of
mental agony—he strives with all his might
to get back again and when he succeeds,
he plunges into its middle, to bathe
his whole body with the comfort
of close companionship.

—Francis Galton, 1871[1]

ONE OF THE GREAT MYSTERIES of human history is how and when animals became domesticated. A rather grey haze settles over the details of the first herding, breeding, castrating, and butchering of animals. Popular myth has it that people herded first, then learned to settle down and farm fields, yielding larger amounts of food that could feed larger numbers of people. But that is not the way it happened. Researchers have evolved an explanation for the domestication of herd animals, particularly cattle, that goes about it in a different manner.

From archaeological evidence it is certain that cattle had been domesticated in the Fertile Crescent area of Mesopotamia by 4000 B.C., but nothing is known about these beginnings. Cattle, or *Bos*, the family of bovines, had spread out over Europe and Asia after the Ice Age and had moved into all the temperate zones of the Old World. That wide dispersion is what makes it impossible to pin down exactly where they first became domesticated. Cattle are definitely an Old World animal; no remains of *Bos* have ever been discovered in the New World. American bison, a bovine relative, roamed the grasslands of North America, but until the coming of the Spanish explorers there were no cattle in the New World.

According to Carl Sauer, the myth that hunters became nomadic herders, then pastoralists, then settled down as farmers has no basis in fact. Hunters were not animal domesticators. The idea that hunters began herding wild animals into enclosures, thereby isolating and holding them, which somehow led to taming and domestication, is another erroneous idea, he insists. Early people did not have the technology to build secure structures to hold aurochs, the cattle of the time, which were large and wild. Nor could wild cattle have been starved into submission while held in an enclosure. People who adopted agriculture were ax users, not hunters, who lived in the woodlands along waterways. They were sedentary, their diets made up of small game, snails, shellfish, and nuts. Life was easy; the only effort was making seasonal trips to gather. When animals such as cattle and goats began watering and feeding nearby, they too became part of the people's food supply.[2]

Another explanation for animal domestication revolves around the idea that humans tamed wild animals as pets, which grew into herds. Some theorists propose that a wild baby animal was captured and reared by humans, and was viewed as such a success that it quickly resulted in others taming animals—and herds developed. Not likely. People all over the world, in many cultures, capture wild animals as pets. But that has not led to their taming or domestication. Cattle came from two wild breeds, *Bos primigenius* in the West, from Atlantic Europe to western Asia, and *Bos namadicus* in India. Both *Bos* types were present in Indo-European cultures, and both were eventually domesticated, but not because they made good pets.

However it originated, cattle herding developed into domestication, which then filtered out from Mesopotamia to the length and breadth of Europe as well as the length of Africa. People living on the outskirts of large Indo-European agricultural settlements, such as the Fertile Crescent of Mesopotamia, initially developed milk and beef industries that spread to northern Europe and east Africa. Cattle milking began in southwest Asia but was never established in southeast Asia or the Pacific Rim.

Milking cows probably began as a religious ceremony, not an economic activity. While we tend to see historical developments as economically driven, it is not always so. Milk, for example, was used as a ceremonial sacrifice within the maternal societies of the ancient West. In societies directed by queens, priestesses, and prophetesses, milk and its products (curds, butter, cheese) appear to have been used as ceremonial offerings. Animal sacrifice developed later, as society became more patriarchal. Domestic animals were perfect for religious sacrifice. Early on, the sacrificial meat went to priests and the elite; eventually the commoners were allowed to participate too, and the eating of meat became available to all. The animal was initially sacred and became common, not the other way around.[3] Beef never did become an ordinary meat in the Near East or the Mediterranean.

Pastoralism, the raising of small herds of cattle, sheep, or goats, began on the outskirts of farming communities in the rich Indo-European centers of early agriculture. Herds grazed in areas outlying the tilled fields, and as the numbers of animals increased, additional pasture had to be hacked from the forest. Axes were part of early cattle domestication, useful in hacking away brush to create pasture; the double-bladed ax appears significantly as a decorative and symbolic element in early cattle cultures. As the new grazing areas were cleared, the herds drifted farther away from the population center until eventually they and their people detached completely from village life and began to live as nomads. Still they needed grain grown by their farming neighbors for cattle feed, and pastoral groups developed out of these farming cultures where livestock was already present.[4] When the people began moving with the herds, in some ways one could say that the people became members of the cattle herds, both relying on sedentary farmers for supplemental grain supplies.

The earliest clearing of land was not for fields but for pastures and meadows. Wild plants sprang up in the new ecological niches, filling the cleared pastures with two plants cattle found irresistible: rye and oats. They grew well in acid soil with lots of moisture, the perfect crop for newly cleared forestland. Wheat and barley languished in such soil and climate. Eventually rye and oats became the cereal staples of northern Europe and Britain, but at first they were encouraged as cattle feed. Farming in northern Europe was a combination of raising seed grains, clearing new ground for pasture, and cutting hay. This provided the feed for cattle herds that were put into dairy production. Hay and grain could be set aside for use in winter, resulting in the development of protective structures: barns. Stables were added, making the handling of the large animals easier. The hay crop was actually more important than field crops. It enabled the cattle to survive and fatten, vital to the people's continued existence.

As Frederick Zeuner writes in *A History of Domesticated Animals,* the trait that "distinguishes man from all other animals is that he, instead of adapting himself increasingly to his environment, has undertaken to subdue the environment to his purposes."[5] This economic switch—as people avidly cleared more pasture land for their animals of choice—marks the start of an all-important effort by humans to interfere with their environment. And by "domesticating" the environment—clearing forests, planting feed grains, cutting hay, and building barns—people were able to domesticate cattle. It was through manipulating the natural world that they were able to maintain their control over cattle.

Domestication was inevitable, according to Zeuner, because it was part of the ongoing process set in motion by people's habits and the biological makeup of certain animal species. All animals that were eventually domesticated (and no new ones have been added since 2500 B.C.) had a social life of some kind while in the wild state, either in packs or herds.[6] Animals that naturally interact on a social level with other members of their species will also interact with other species. Mixed herds graze easily together in a natural setting. With people and cattle, it was a symbiosis—both species gained an advantage from living together. Cattle obtained adequate food supplies and protection from predators, because humans provided both. For humans, cattle provided a future food supply that could be stored "on the hoof," at a time when food preservation was difficult. They also constituted a renewable source of wealth that increased each year.

Once the cattle were accustomed to domestication, they were fairly easy to raise in confinement. Their breeding was controlled by destroying or driving off wild cattle to prevent interbreeding which might result in undesirable traits. By focusing on the traits people wanted to see continued in their herds, it was easy to breed smaller, tamer, animals. Domesticated animals did become smaller—except for birds and the horse, which were bred to larger sizes. Cattle colorations became rather set: bulls were usually black with a cream-colored stripe down the back (just like many of the cave paintings) while cows tended to a reddish color.

So it went in Europe for two to three thousand years. After Columbus crossed to the New World, new plants were brought back to the Old World and cultivated, but the new acquisitions, particularly potatoes and corn, were first fed to cattle. Only later were they accepted into the human diet. By the eighteenth century, beets, turnips, and clover were introduced into Britain and western Europe for cattle feeding.

Early farmers grew plants that yielded high levels of starch and sugar but inadequate amounts of protein. That was fine for cattle, whose digestive system utilizes starches and sugars to synthesize amino acids. For people, however, a diet that shifted from hunting and gathering to crop consumption was deficient. Plants, particularly grains, simply do not yield as much dietary protein as do animal sources. And plant proteins lack essential amino acids, requiring the addition of milk and eggs to a diet that is low in meat or fish. As a consequence, many early agricultural peoples suffered from inadequate diets, usually deficient in protein and fats. Unless supplemented by hunting and fishing, adequate nutrition was impossible to obtain. Anthropologists have examined skeletal remains from early European peoples, living 30,000 years ago, and found them to be tall, in fact perhaps taller than we are today. Their descendants, who relied on agricultural crops for nearly 90 percent of dietary needs, were so deficient in protein that clinical malnutrition was characteristic, evident in the fossil remains. Before European contact with the New World, Old World farmers eked out a hard living raising grains, on a diet that weakened them and made them even more susceptible to the infectious diseases that were rampant in densely populated towns and cities. Not until after the Industrial Revolution did Western diets improve enough to show an increase in the average height of Americans and Europeans. Our height has increased in the last 150 years; we are now about as tall as Paleolithic people.[7]

Crop-growing cultures displayed other characteristics not found in pastoral people. They developed cultural dietary practices utilizing yeasts (to make bread rise) as well as fermentation to create alcoholic drinks. Fermented crops led to the making of fermented drinks, something cattle cultures never adapted to. Beer, wine, alcoholic drinks—all are the product of farming cultures. Alcoholism has been postulated to be more common among cattle cultures, which have not experienced centuries of adaptation to alcoholic beverages. It is indeed more common among people whose ancestors did not raise grain crops: Scandinavians, Irish, Scots, and American Indians. Another dark side to plant-based diets is that cannibalism is a practice singularly found in planting cultures, where fresh red meat is nonexistent. Examples have been found in Southeast Asia, Africa, and Central America. The Spanish explorers to the New World, who brought cattle with them on their ships to ensure a fresh supply of beef, were shocked to encounter cannibalism in the Indians of the Caribbean. The word *cannibal* is derived from *Carib*, the name of the Indians who prepared "man stew." The Aztecs made human sacrifices, and the consumption of the body parts afterward was central to their ritual and society; flesh was distributed to the elites for consumption in their homes.[8] Native American groups that relied heavily on corn, such as the Iroquois, also developed ritual cannibalism. Plains Indians, however, whose diet consisted of little besides fresh meat, seldom participated in any cannibalism rites in spite of the aggressive, violent, and often bloody culture they lived in.

Even in the early stages of domestication, cattle were valued for more than simply steak. The ancient Sumerians had a well-developed cattle culture, and by 3200 B.C. their cattle were being used to pull implements for field agriculture. Clay cylinder seals depict cattle scenes: herdsmen feeding them, men driving a bull and herding cows and calves, and a goddess being pulled on a sledge by oxen. From the temple at Ur a wall frieze depicts a scene of men milking cows from the rear—goat-fashion. Sheep and goats were milked from behind, so this image shows that cattle milking must have followed after the adoption of milking goats and sheep. (Milking a cow from behind must have been a highly distasteful procedure. Cows relax when they let down their milk, and that does not bode well for anyone positioned beneath their tail!)

Oddly, images from Sumer always show cattle with one horn removed—the right one—and even indications of a scar on the animal's head where it had been removed. One explanation may be that cattle were ridden into battle, the right horn's removal making it easier for a right-handed warrior to throw spears or shoot arrows without the animal's horn interfering.[9] Like the women warriors of Scythia, who had their right breasts removed in order better to use a bow and arrow, cattle may have been at least ceremonially altered for a similar reason.

In Egypt, too, public art is replete with images of cattle, but evidence of domestication is scant. Wild cattle roamed the Nile Valley swamplands in pre-dynastic times, and hunting them in horse-drawn chariots was popular among Egyptian elites. Tomb reliefs depict cattle pulling plows, tramping grain on threshing floors, and being milked. Men are shown roping and throwing a bull, and slaughter scenes detail the careful, methodical butchering of the carcass. Grazing cattle accompanied by herdsmen who tote bedrolls on their shoulder, ready to sleep with their herds at night, show everyday scenes of pastoral life. Scenes also depict a procession of Nubian princes to the Egyptian court: a long procession of cattle includes princes in a chariot pulled by bulls harnessed with bits in their mouths, adorned with rich trappings and ornamented caps on their heads. Other Egyptian scenes show cattle held in carved wooden cages being transported on the Nile in open boats.[10] Cattle were a significant part of the economy; Egyptians took a cattle census over the entire country every two years so that wealth and taxes could be ascertained.[11] Cattle were also prominent in Egyptian religious beliefs. The bull appeared in ancient Egyptian religious artifacts, and mummified remains of bulls have been found in sarcophagi. Hathor, the sacred cow, had a human head with cow's horns.

Cattle spread from Egypt into other eastern African grazing regions. The Dinka and the Masai are the best-known cattle cultures in Africa. The Dinka women, in the Sudan, have long performed a dance with their arms outstretched and hands pointing upward, like bull's horns. The dance is called the Cow Dance and is done simply because cows are "pleasant, useful creatures," Zeuner explains. Dinka cattle herders also mutilate one horn on their cattle, perhaps related to the long-ago Sumerian practice shown by one-horned animals in Egyptian and Mesopotamian art.

No discussion of cattle images in early art is complete without mention of the Cult of the Bull which developed in Minoan Crete. Sports and contests in which acrobats performed with cattle were hugely popular. The cattle contests were likely the result of people trying to catch nearly wild cattle. Their role went far beyond simple food source. The tale of the Minotaur, a half-man, half-bull beast that devoured virgins, sprang up before 1600 B.C., when Crete was a major sea power in the area and when annual sacrificial rites featured humans pitted against bulls in a survival contest. The Minotaur was the hybrid offspring of Pasiphae, wife of King Minos, and a bull. Pasiphae was a nymphomaniac, according to legend, who sought a bull for satisfaction. Daedalus, an engineer employed by her husband, built a fake cow in which Pasiphae could hide while she mated with a bull. The legendary coupling caused her to give birth to the Minotaur, a societal misfit, who was placed in the center of a labyrinth from which he could never find an escape.

The Knossos Palace in Crete was a site for bull-sport performances. Large audiences assembled to watch a band of unarmed male and female acrobats seize the bull by the horns, so as to be thrown up over the bull's back in a somersault maneuver. Minoans worshiped the bull, but this performance appears to have been only for spectacle and thrills. The palace was decorated with ornamental cattle heads and horns on the walls, over doorways, and throughout the structure. Often the images of cattle heads and horns were accompanied by the double-edged ax, the symbol that had come from Asia with the cattle.

THE DOMESTICATION OF CATTLE was an important step for civilization—economically, socially, and politically. Cattle require attention and a community that is sufficiently organized to ensure that the animals are cared for and fed, and that the meat is distributed fairly. Rules about slaughtering the animals must be enforced in order to promote the propagation of the herd. One animal, butchered before food-preservation techniques went beyond drying, may have set the parameters for group size. How many people could eat one beef? That may have been the optimal tribe size. A group needed enough members to herd, feed, and handle the herd, yet if there were too many people, there would not be enough meat to go around. Cattle's large size and strength meant people had to solve problems, including how to secure and handle the animals as well as how to butcher and preserve the meat. As cattle culture developed, knowledge, technology, and planning were required to supply and store adequate feed for winter. Not only did the people have to survive, so did their animals.

The number of cattle kept by early herders was small; the people still relied on hunting for much of their meat. As in China and Egypt, cattle were usually worth much more to humans alive than dead. People maintained cattle to ensure a steady supply of milk and blood: not only did the Masai periodically harvest fresh blood from their cattle, Scots and Irish had done the same thing. Cattle dung provided another valuable commodity. In areas where fuel was scant, dried dung served quite adequately for cooking and heating. Nineteenth-century pioneers on North America's prairies gathered "buffalo chips" and cattle dung from their own livestock to use as fuel. Some African groups built houses from cattle dung. But the most important asset cattle had was their strength. Their ability to power wagons, sledges, grinding stones, and grain threshing equipment was invaluable. Oxen were male animals that had been castrated (easier accomplished while they were young calves), which made them more docile and agile than bulls. Stronger, calmer, and easier to handle than a horse, oxen provided the power for humankind's greatest invention: the wheel. For millennia, humans relied on oxen to power civilization over much of the world.[12]

STEPHEN BUDIANSKY, a journalist living on a small farm in Maryland, has a different view of the relationship between humans and livestock. In his book *The Covenant of the Wild: Why Animals Chose Domestication,* Budiansky argues that our contemporary view of animals has been clouded by socio-political explanations. Rather than seeing farming as "exploitation" of animals, as animal rights activists are prone to do today, he thinks domestication is a process crafted by forces we have never been in control of.

Until the Industrial Revolution, Budiansky points out, animals were treated as animals. They were respected for their own natural attributes, whether wild or domestic. But with the onset of industry, people began to view animals as "machines," suitable for use and then abandonment. Urbanization distanced people from agricultural and pastoral life. As animals became badly treated, a backlash set in. A romantic vision of animals that were adored for their "lovableness"—such as Black Beauty, Lassie, even Elsie, the Borden Milk cow—captured popular imagination.

While animals have become icons, we have little understanding of their natural life because we never experience or interact with them. Pets, with their human attributions, are valued for the ways they resemble *us,* not their wild ancestors. Today we use animals as symbols of a long-lost Eden, a connection to a religious "wild." Yet we continue to depend on animals in many of the same ways as our distant ancestors did. We rely on them for food, medical uses, and increasingly for vital pharmaceuticals and nutritional supplements.

We have also created a false explanation of animal domestication. Budiansky cautions that we are so "accustomed to the notion that domestication was a human exploit that to suggest otherwise can make one sound more like a mystic than a scientist."[13] He argues that domestication evolved from a mutual strategy for survival that developed between humans and animals. Domesticated plants and animals were really interlopers seeking an opportunity for easier survival. The first animals that were "domesticated" were social species which scavenged for their food: dog, sheep, and cattle.

During the period that saw the rise of agriculture, humans, plants, and animals slowly developed interrelationships. They survived in greater numbers once they had become dependent upon one another. This dependence flies in the face of evolutionary thought, driven since the era of Darwin, which sees survival as a competitive rather than a cooperative venture. But nature is more often a web of cooperation, Budiansky argues, than a competition between species.

While we view our ten-thousand-year-old relationship with livestock as a product of our own ingenuity, we pay little attention to our dependence upon the animal world. While we cannot ignore the importance of human inventions, as a species are we in command? Budiansky claims that, "in an evolutionary sense, domesticated animals chose us as much as we chose them."[14] What if we had never been asked to dance? Would we still be sitting on the sidelines? Jared Diamond explores that very question in his book *Guns, Germs, and Steel,* in which he points out that only a few areas in the world had the "perfect" animals to domesticate. Only a few species—dogs, horses, cattle, sheep, goats, rabbits, some birds—that could *ever* be raised in captivity. Once those animals had located next to people's settlements (all were scavengers, "crop robbers"), they and their new masters began adapting to one another.

We cannot claim that animals deliberately sought out humans and positioned themselves near them; the dynamics of environment had much to do with it. In many parts of the world, archaeological evidence shows that farming coexisted with foraging and hunting for thousands of years. The farming revolution may have been a slow response to the environment, an evolution in behavior. Along with understanding soils, climate, and the requirements of plant growth, people began to understand the behavior of the animals they could not seem to get rid of. We assume that all animals want to escape from humans and flee to the wild. We do not notice that mice invade our homes, deer graze impetuously in our backyards, and now, across the West, cougars circle playgrounds and parks. Coyotes, raccoons, and others sneak into our yards in spite of our best attempts to keep them at bay with repellants, gimmicks, and fences. Was it the same for early gatherers? Were cattle competing for favored grains and plant foods? What is natural behavior and what is man-made?

As the Ice Age melted away, a few species were able to form relationships with humans that have brought them to a dominant position on the planet. Dogs, sheep, goats, cattle, and horses vastly outnumber wild animals now. Coinciding with an increase in human population, these animals increased in vast numbers after domestication. Animals that fearlessly approached man, and allowed people to enter their herds, survived; many other species faded into extinction.[15]

It was not so much that domestic animals began to behave unnaturally; instead it was their particular disposition, their instinctual makeup that made them well suited to live alongside humans. They were dependent, cooperative, opportunistic (they *were* scavengers), curious, and fearless. Their social structure was part of it too. Cattle have a social hierarchy based on dominance and submission. Without such a system to shape behavior, herd animals would spend all their time fighting and have no time for mating, rearing young, or foraging for food. The dominance-submission system provides a simple framework for group living. Things go along smoothly, everyone knows the rules, and everyone survives. Dominance is achieved through initial combat; and in some cases animals simply submit to an animal that exhibits confidence or a mildly threatening manner. Cattle live in groups, and when humans are added to the group the hierarchy changes to include human dominance while retaining the basic social structure the animals have always had. People have simply been added to the herd.

WHAT MOST DISTINGUISHES animals that have become livestock is their social behavior as members of a herd. For the most part, herd formation is a defensive behavior. A group of animals, females and youth to the center, are safe because a few individuals on the perimeter act as lookouts for predators in the area. The dynamics of the herd serve cattle well. One of my neighbors, Hugh Williams, points out that cattle individually are quite stupid, but the herd itself is incredibly intelligent. His herd of about sixty animals operates as one, communicating with one another by small noises, a sense of hearing, and a sense of body movements. Because they follow a leader to new pastures, they spend less time looking about, wondering what to do. Once a leader is established (and for cattle that tends to be a dominant cow, who thinks independently), the others do not waste time and energy arguing or balking at going along with the herd. In Hugh's herd, which is mostly Scottish Highland and some Belted Galloways—both ancient breeds—when the lead cow is not with the herd because he has put her elsewhere for some reason, her daughter takes over in what can be described as a hereditary position of authority. The bulls rarely lead the group anywhere, the lead cow does that; but the bulls remain at the rear of the herd, pushing along the stragglers. In that position they provide protection for the weak and aged members of the herd from any predators that may be following the group. It is also an advantageous position for them, as the herd acts as a buffer between them and predators that they may encounter as they move.

When a cattle herd moves of its own choice, the high-ranking cows will take the lead. But when forced to move, such as when driven by herdsmen, the herd pushes the low-ranking animals to the front. This use of marginal members of the group as a shield against the unforeseen is a social way of avoiding predators. Because predators tend to grab the first animal they encounter, it pays for animals to seek the center of the group.[16]

This centripetal movement toward the center of the herd was first identified by Francis Galton, during a year he spent living among the cattle cultures of western South Africa in the 1850s. Galton, the cousin of Charles Darwin and an eminent English scholar, later published *Hereditary Genius,* a book that laid the foundations for eugenics, the social manipulation of human genetics through selective breeding. Traveling and living with the Damara people in Africa, along with their hundred or so cattle, Galton became intrigued by the behavior of the animals. He rode them, drove them, ate them, and at night, "I lay down in their midst," where he observed how the cattle moved in to sleep near the camp's fires, "conscious of their protection from prowling carnivora, whose cries and roars now distant, now near, continually broke on the stillness."[17]

Galton, a pioneer in the study of anthropology, heredity, and eugenics, was fascinated by the behavior of cattle. For him, "the habits of the animals strongly attracted my curiosity. The better I understood them, the more complex and worthy of study did their minds appear to me."[18] He found the animals to be self-absorbed, even while in the middle of the herd. And while they showed little fondness for one another and generally remained detached, when one of the animals was unwillingly separated from the herd, "he exhibits every kind of mental agony; he strives with all his might to get back again, and when he succeeds, he plunges into its middle, to bathe his whole body with the comfort of closest companionship."[19]

Their reluctance to take the lead when being driven by herdsmen made good lead oxen particularly valuable. A lead animal or team was necessary so that others would follow, but none wanted the insecure position in the front. Trying to move cattle was difficult as the animals jockeyed around one another trying to avoid being in the lead. Galton describes trying to get a herd moving as being like a "host to a company of bashful gentlemen, at the time when he is trying to get them to move from the drawing-room to the dinner-table, and no one will go first, but everyone backs and gives place to his neighbor."[20]

Galton felt that lead oxen were "born leaders" and that ordinary animals were suited to follow, inevitably to slaughter. "But the born leaders are far too rare to be used for any less distinguished service than that which they alone are capable of fulfilling." Out of any population, human or cattle, only a small percentage would be fit to lead.

Galton realized that cattle had to spend most of their day with their head stuck in grass, feeding, and several hours ruminating in order to digest the grass; they had few opportunities to be alert for predators. They were prisoners of their huge bulk, their digestive system, and their need for a great number of calories from plants. For them, living in a herd provided tremendous benefits. Each animal did not waste energy being continually alert. Cattle in a herd "received a maximum of security at the cost of a minimum of restlessness," Galton reasoned.[21]

Their herd behavior protected them but also made them the perfect domestic animal: they did not frighten and dart off, and their herd habits could be understood and used to control them. A large, fast-growing animal with a curious rather than a nervous disposition, and that ate plants humans did not, was a boon to humanity. They grew and multiplied naturally, were mobile, could be traded easily, and if necessary could be liquidated as meat. They were the perfect form of wealth. *Pecunia,* Latin for money, is rooted in *pecus,* meaning cattle. The word *capital* derives from *chattel,* based on cattle. Owning cattle would become the main objective of many people's lives.

3

Cattle and the Clan

He bringeth forth grass for the cattle: and
green herb for the service of men.
—Book of Common Prayer, 104:14

When we think of small bands of hunting and gathering people, we envision an idyllic society that is nearly problem-free—as close to Eden as it gets. Gathering nuts and berries, a few fish perhaps, is a scene we fill with images of happy bands of relatives popping berries into one another's baskets, chattering gleefully while picking the fruits of the earth. What could divide them? No raids from neighboring tribes, no crop disasters to worry over, no fields to toil in. They would be smiling, gentle, enjoying life one day at a time.

Napoleon Chagnon, an anthropologist at the University of California at Santa Barbara, studied a modern sort of hunting and gathering people, the Yanomamo living in the Amazon rain forest, and found them to be some of the most violent people alive. They fought incessantly over transgressions, and 30 percent of adult males died from violence.[1] Archaeological studies, too, support the findings that hunter-gatherer people lived quite violent lives. The skeleton of Kennewick Man, a recent find in eastern Washington state, has a spear point firmly embedded in its hipbone, a remnant from a violent encounter with the most dangerous animal of all: other humans. Other finds show that early people often died at the hands of each other, and skeletal remains show marks of violent death. Some finds reveal cannibalism, particularly of women and children. Certainly early people lived difficult and dangerous lives, and many died during perilous hunts or from disease, but remains also show many instances of injury inflicted by Stone Age axes, and bones stripped of their flesh by stone knives.

A case can be made that humans became less violent when they began to live with domesticated animals. They became caretakers, nurturers, even fond of animals. People whose survival and security depended upon a herd had a future beyond this hunting season or the next. They were in it for the long haul, for the lifetime of the animals as they watched generation after generation of cattle and sheep emerge in their care. Not only did livestock make people rich, they made people settle down and plan for the future.

Just as cattle found freedom from predators and starvation by mingling with humans, people benefited from cattle too. Clans were a logical development in creating a system to herd and care for livestock as well as slaughter and butcher them. Cattle became a commodity of value, giving an entirely new dimension to people's lives. Property—which is what cattle were, long before land became important—brought social hierarchy, cultural practices, caretaking responsibility, and an interest in progeny who would inherit the herd.

People could move beyond attacking one another over personal grievances, as bands of hunter-gatherers like the Yanomamo did. They banded together, herdlike themselves, to follow an independent-minded, confident leader. The new herd, made up of the cattle and the clan, now moved in sync. Warriors developed to protect the herd from the neighbors as well as to raid the neighbors' herds. Raiding at least took the violence out of the family and the clan itself, and focused it in a positive manner: obtaining more cattle by taking someone else's. The people no longer preyed on one another but formed clans for protection. They no longer faced starvation, either, because they had food supplies secured for many seasons to come—in the cattle herd itself.

Donald Worster, a leading environmental historian, emphasizes how living nature works by the principle of interdependency. "Indeed, it can only work by that principle," Worster points out. "No organism or species of organism has any chance of surviving without the aid of others."[2] We all need evolutionary companions. Civilization and its changes have only been altering patterns of interdependency, Worster claims. Models of successful adaptation are all around us; we just have to notice them. There is only one commonality among all the communities in nature that have survived to this point: they made rules, lots of them. Worster describes these rules as what allowed the communities to govern their behavior, which in turn helped them survive. Having rules and enforcing them vigorously seems to have been essential for long-term ecological survival, Worster argues.[3] Rules and their enforcement had to be developed in order to adapt human families into the clan system that centered on the keeping of cattle and the nomadic pastoral lifestyle that evolved from it. Decisions about what and how to feed cattle, which ones to keep or trade, which to breed, and finally whether to eat them or not, all called for the development of rules.

Cattle culture has been extremely important in the growth of Europe, particularly the northern and western regions. But it is difficult to pin down exactly when cattle-raising began in Europe because we have no accurate way to identify the practice, nor enough remains or explanation. Human bones from early sites in central and southern England show that in Britain early people were eating animal meat and by-products, such as milk and cheese, and that plant foods were of little importance in their diet. Bones dating from the Neolithic period (4100 to 2000 B.C.) have been tested by measuring the stable isotopes in bone protein. If human bone values are similar to those of herbivores (horses or cattle), the people ate pretty much a vegetarian diet; if the isotope levels in the human bone are more like those of carnivores (wolves), the people were eating a diet of mostly meat. The tests showed that the isotope levels of these early people were as high as carnivores, suggesting they were eating a lot of meat, possibly including milk and cheese (those isotopes are indistinguishable from that of meat). Their economy was based on domestic animals; we know that from the archaeological remains found at their home sites. Their cattle were well kept, and remains show them to be larger than typical Iron Age cattle. Findings suggest they were an animal-dependent culture which treated their animals well and kept them a long time. The British Stone Age people may have been, as Neolithic specialist Andrew Sherratt suggests, part of an agricultural revolution marked not by the growing of crops but by the use of animal dairy products. Grain and farming implements have also been found at the same sites in Britain. Archaeologists suggest the grain must have been grown for ritual purposes, but its value as animal feed during long winters may have been more likely.[4]

Stonehenge, one of Britain and the world's most mysterious structures, was made during this period of time. Burials in and around the upraised stones include cattle skulls and bones buried along with human remains. And if the builders moved the stones into position using sledges and log rollers, they certainly would have put oxen to work as draft animals.

The hazy early days of cattle and clan are difficult to explain, so we turn to the history we do have, besides highly technological tests of protein or carbon remains. We can look at folklore and mythology that have been passed down through the ages, over campfires and feasts, during rainy seasons and long dark winters, long before they were put down in writing.

The story of Cain and Abel (Gen. 4:1–16) is in its origins a folktale that seeks to explain how men began to worship God. It is rooted in the nomadic life of ancient times when people were trying to determine what sort of sacrifice God preferred. Cain's sacrifice—the crop from his field—was found inadequate when compared to his brother Abel's sacrifice from his flock. At that time animal sacrifice was still preferred over agriculture; it would be centuries before the Christian church adopted the use of bread as sacrificial food. Genesis, told from a pastoral point of view, points out the superiority of animal sacrifice over field crops, at least in God's eyes. But the story also deals with brotherhood on a different level. Cain asks the question, "Am I my brother's keeper?" expecting to be let off the hook. God answers that indeed he is, thereby defining brotherhood's limits which have been defined over time as family, clan, nation. Cain, meanwhile, jealous and envious of his brother's favored sacrifice, reacts with the ultimate human sin: he kills his brother. He is doomed forever as a sinner and an outcast.[5] Abel was not a cattleman, nevertheless the story points up the value that ancient peoples placed on animals.

In cattle cultures the herd has economic importance, is the measure of wealth as well as a food source. Besides food, it gives other products such as hides, bones, and dung. Cattle are seen as the source of goods in the cultures that depend upon them. But cattle also play a role in social transactions. Their value translates into human value; bride price has often been determined by number of cattle, a clan's way of limiting births to the group's ability to support more children. And the cost of revenge for a wrong, or blood price, has been paid in cattle because only they are valuable enough to make up for the loss of a human member of society.[6]

Cattle are part of the social order and part of the community of goods and people. Cattle owners hold intense affection for them as well as a desire to own a great many of them. Longing for cattle has resulted in cattle raids on neighbors, whether in a cattle culture of northern Europe or eastern Africa. Warfare has centered on cattle—no other booty will satisfy. According to Bruce Lincoln, in his book *Priests, Warriors and Cattle,* this is because cattle form the basis for economic, social, and religious life. The Dinka of East Africa lovingly care for their cattle, calling themselves "slaves of cattle."[7]

Lincoln points out that religion is rooted in culture, and culture is shaped by environment. When that environment includes cattle herds, the culture shapes religious ideologies.[8] Lincoln describes similarities and connections between Indo-Iranian mythology and that of the Norse, the proto-Germanic people, Romans, Irish, and East Africans—cattle cultures all. He links cattle to the idea of a warrior culture, describing how cattle societies feed their warriors huge amounts of meat to build their strength. The warrior's role is to raid for more cattle, and it assumes a ritual nature, with the involvement of priests, good omens, and charms. Prayers and special rites ensure a successful raid. Raids, according to Lincoln, connect to the creation myths of cattle societies. They explain how a group began by obtaining their first cattle.

In cattle-keeping societies, cattle were important in myths, rites, gifts to gods, and requests of them. Cattle were a constant concern in the daily life of African and Indo-European societies. Raiding for neighbors' cattle seems to come automatically to cattle societies, where the number of cattle kept becomes vastly important to the group. Raiding gains more of the prized commodity. A sacrificial aspect relates to the way this valuable commodity is shared: how the meat is distributed. This sharing is rooted in the very essence of brotherhood and sisterhood. As clans became more sophisticated, a priest class emerged, whose job was to make sure the wealth (meat) was distributed appropriately: some for the gods, the rest among deserving members of the group.

Craig Stanford, an anthropologist who has studied how apes and monkeys distribute meat within a society, explores this division of the spoils. In *The Hunting Apes: Meat Eating and the Origins of Human Behavior,* Stanford argues that the origins of human intelligence are linked to the acquisition of meat and the cognitive skills necessary to share the meat. Meat sharing means deciding who in the group gets meat, how much, and how often. It also determines who is refused or ignored, and why. Stanford believes that this thinking about how to share meat led to the expansion of the human brain.[9]

In the 1960s, anthropologists began to promote the belief that early people ate diets composed mostly of plant materials, and began to minimize the importance of meat in early diets. It had been customary to think of men as the central players, with much attention to mating rituals and the theory of "Man the Hunter." The scholarly backlash placed more emphasis on the role of women who foraged for nuts and small animals. New thinking now places females at the center of the mating system too.[10] Certainly women (and children) obtained large amounts of dietary protein from small mammals and plant foods. Nevertheless the place of meat in society cannot be minimized. Even when it did not make up the bulk of the diet, it was extremely important. Looking at primates, Stanford was intrigued by their systems for sharing meat—its use as a political tool, a social manipulative, a reward for allies, a punishment for rivals (if withheld), or a gift to obtain a mate.

Meat sharing was important in the development of rules and behavior in hunting societies, where it laid the foundation for manners as well as politics and government in the cattle cultures that followed. Because a successful hunt requires several people and their cooperation, a system must ensure that all available hunters take part. No shirkers can be allowed to loiter around the campfire, eager to eat the spoils of someone else's work. Instead, all must take part in order to ensure a sense of justice within the group as well as a successful hunt. The only way to ensure that everyone participates is by not sharing meat with those who do not hunt. Meat sharing becomes a political tool that contributes to a smooth-running society. And, the sharing of surplus meat can even put individuals in positions of esteem as well as garner them willing mates. Researchers have discovered that males who brought back and shared large amounts of meat with females were most sought after as mates. Seen as both good hunters and generous providers, such men were quickly accepted, even pursued by females.

Stanford describes the concept of "Machiavellian intelligence": primates (including man) who could best manipulate their social surroundings to their advantage reaped more mates and left more offspring (or were successful by today's standards). Political leaders have used this theory to justify their dominance throughout human history. Leadership evolved not to the best hunter or warrior but to the one who was most politically astute. High cognitive skills, which made the best use of the meat distributed after a hunt or cattle raid, determined leadership. Stanford points out that in judging the sophistication of a social group it is not the behavior during the hunt or raid that is significant, but how they behave

WITHDRAWN

while dividing the spoils.[11] Coalitions of people for the purpose of capturing game animals in a planned hunt or by raiding for and raising cattle, produced egalitarian groupings that could easily challenge those who tried to form a dominant hierarchy. Whether it was a king who laid claim to the land beneath them, or a church that sought to control their lives, people in a cattle culture were likely to join together at a lower level to challenge authority. They were used to working together, forming coalitions and alliances, and distributing authority. It made them very difficult for authorities to "tame," and these age-old traditions and ideals have resisted change. Scots, Irish, and Norse—bastions of Celtic cattle cultures—were difficult for their neighbors to control. In Africa, the cattle cultures continue to live in conflict with authorities who seek to settle and contain them. Cattle people are not easy to conquer or to control because they cooperate at lower levels of resistance, a trait developed in the egalitarian clan where everyone cared for and had a stake in the cattle.

A society based on the division of spoils is by necessity an egalitarian culture. In hunter-gatherer societies there is no top-ranking male who dictates what the group does or where it goes. Successful hunting societies and cattle societies have the same element at the center of their social world: the distribution of animal products. Like the hunt, the herd is a shared endeavor; the act of herding is not a one-man show. Cooperation is imperative at all levels, including the distribution of meat. Clan leadership exists, indeed it is important to the success and survival of the clan; but that leadership is tempered by the membership below. How the clan leader shares the bounty of the raids, how the leader rewards or punishes individuals by giving or denying them cattle of their own, or hides, or meat, or whatever valuable is under consideration, determines whether he will remain in a position of leadership. Consensus is all.

In this egalitarianism, individual interests are sometimes submerged in order to benefit the group, or clan. Consensus leads members to "do the right thing." The goal of the group members is not to get ahead for oneself but to ensure that no one else gets ahead of the rest of the group. This attitude prevents anyone from greedily taking advantage of the group's property. No one is allowed simply to butcher and eat the cattle at will. They belong to the whole in many ways, even if considered the property of a few individuals. When groups began to share common pastures in Europe and later extended that practice into colonial New England, they made certain that no one individual took advantage by grazing more animals on the commons than others did. Careful structuring of the grazing arrangements, as well as the distribution of winter hay and feed stores, prevented any single individual from manipulating the group's interests for himself.

But things changed with the onset of the Industrial Revolution, the rise of imperialism, and the growing urbanization of societies. Social castes and institutions based on socio-economic status developed in response to specialization. These systems are male-dominated, patriarchal systems where males control the assets, dispensing them at their choosing. In urban societies where people seldom own and control property, controlling resources means controlling the population. People may own their home, but it represents a savings account for retirement rather than something to pass on to the next generation. The stock market may be today's symbol for the herd; if we pay attention to stocks, they should show a natural increase. We even call it a "bull" market when stocks increase—a metaphor for breeding more cattle.

Cattle cultures were more egalitarian and largely female-dominated, or at least gender neutral. Women have always been important players in the domestication of herd animals, particularly those used for dairy products. In cattle cultures, oxen were used for draft work, something often supervised by the children of the family. The cows, who may have worked the fields as well, were in the care of the females of the family, who did the milking and processed the milk and cream for home consumption, barter, or sale. Calving, calf-rearing, animal medical care, and milking were all the domain of the women of the clan or house. Men usually performed the physically difficult tasks, such as hay harvesting and storage, building fences and corrals, and clearing new pastures. Everyone in the family was involved with the cattle; everyone took part in the decision-making about them. The prevalence of women and children in folktales and myth involving cattle show how significant this was.

Place-bound, crop-growing cultures that lacked mobility were easy to control. They were dominated by political and religious hierarchies; decisions about land ownership, taxes, and irrigation practices were easily pushed into a political hierarchy that was usually patriarchal. Cattle-raising cousins, on the other hand, could sometimes pick up and go to greener pastures, and they did. Archaeological and historical records note a continual movement onto new lands by people who took along their cattle. Early Viking settlements in North America brought with them large numbers of cattle and consumed a largely dairy diet. Indo-European people moved onto the British Isles in small hide boats, taking cattle with them. So, even if one could not walk the animals the entire way, owning cattle allowed a clan to pick up and go elsewhere. All that was needed was grass.

THE INFLUENCE OF CATTLE is pervasive in antiquity; their presence as tomb decorations, in economic records, and as part of ancient oral traditions is clear to see. Their influence on how we record written text—as prominent symbols representing language and ideas—is less well known. The alphabet emerged in the Fertile Crescent, in Sumer, at the time cattle were being domesticated. At first it was a system of writing based on phonetic sounds. Devised by the Canaanites, a Semitic people in the Near East, the alphabet today is much the same as the original created more than 3,500 years ago. Because cattle were so influential in these ancient cultures, it is no surprise that the first letter of the alphabet in the Phoenician, Cretan, Greek, and Latin alphabets symbolized the head of an ox—the foremost source of wealth. The letter "A," called *aleph,* represents an ox head turned upside down. The letter B, *beta,* represents a house or tent. The symbols and their position at the front of the alphabet (that word comes from *aleph* and *beta*) signifies how valuable stock animals and pastoral life were.

The phonetic alphabet is the most abstract of all writing systems; it relies on fewer symbols than do other writing systems based on pictographic symbols. This use of abstract sounds to create ideas has led to what has been called "the alphabet effect." The intellectual by-products of the alphabet include abstract thinking, analysis, rationality, and classification—all elements of scientific and logical thinking.[12]

Even the way text proceeds across a page is related to cattle. At first the writing of letters was done in no particular order or direction. Sometimes they were written horizontally, other times vertically. Eventually the idea of writing horizontally across a document evolved, and the writer would turn at the end of the line and reverse his direction, going back across the document in the space below. Called "boustrophedon," that type of writing turned on the page like an ox plowing a field, which is what "boustrophedon" means. By the mid-eleventh century, in Ethiopia and Greece, writing became structured left to right.[13]

Another type of writing, with the *runic* alphabet, developed in northern Europe about the first century B.C. There were several versions of the characters, called *runes,* and their origins are unclear. One theory claims the runes were invented by the Goths, a Celtic people, who got the idea from the Etruscans. Another theory is that the Goths got the idea for the runes after contact with a Greek colony located on the Crimean Sea. Another belief is that runes were invented by a single individual. Today the most popular theory is that they were based on a northern Italian, Etruscan-based alphabet in the first century B.C.[14]

Runes made up a secret writing system used for religious and mystical purposes; markings were made of angles, no curved lines. They were usually carved or etched on wood or bone surfaces—at least those that have lasted until now. Rune markings have been found on stones, coins, and jewelry, and even a few manuscripts. Several varieties were known, the most common being Anglo-Saxon, Scandinavian, and Germanic. Runic alphabets had almost the same number of letters as the Roman-based alphabet, but most of the letters were in a different order, and the script was written from right to left.[15]

Today there has been a revival of interest in runes, at least in divining a sense of meaning from the symbols. A set of runes is usually made of wooden pieces, each with a runic inscription on it. Each symbol is a symbolic representation of objects, actions, or values. To "consult" the runes, one picks a random piece from a bag containing the set of runes. The symbol may mean something to the person at that particular time, but it is not really fortune-telling. Runic readings are most appropriate for self-examination, insight, advice, or a different perspective.

"Reading" runes and interpreting them was not reserved for high holy men; anyone could do it. It was the successful integration of that divination or insight they provided that made one valuable and respected as a seer. At the time the runes developed, northern European cultures lived in small family or tribal clusters, or clans. Their diet and well-being were secure; nature and their cattle provided sustenance for today and the future. They were fiercely courageous and proud, and inspired as much by poetry as by war. The dissemination of knowledge through the runes encouraged people to think for themselves rather than follow a prescribed set of doctrines.

Examining runes and their place in culture can help us understand the values and daily lives of the northern European people who adopted them—who were also cattle clans. Examining Elder Futhark, which is the most popular runic form used today, provides a glimpse into the past. Notice the emphasis on cattle, something in common with the creators of the alphabet.

Here they are in order, an order that has never changed:

1. *Fehu* means cattle. It represents wealth, prosperity, security, goals, and even a good reputation, which was extremely valuable in clan cultures.

2. *Uruz* means aurochs. It stands for the fierce wild cattle that tested one's courage and skill. It also represents raw strength, energy, passion, sexuality, and instinct.

3. *Thurisaz* means giant. This rune is associated with hardship, discipline, knowledge, and force, for attack or defense.

4. *Ansuz* means a god, Odin most often, and represents authority, leadership, and justice—all vital to the survival of a clan.

5. *Raidho* stands for chariot and the act of riding, which means movement, transportation, and travel. It can also represent a person's journey through life, or control.

6. *Kenaz* means fire or torch, and represents enlightenment or inspiration.

7. *Gebo* means gift, and stands for the connections between people. Runic cultures were shared systems, not solitary. A continual message in the runes is that paths connect with other paths creating bonds.

8. *Wunjo* stands for glory, success, reward, and a chance to enjoy one's achievements. It also represents the dynamic of making one's dream or goal come true, not simply by wishing but by actively seeking it.

These eight runes make up the first cycle in the runic system, the first level of understanding. To become a skilled interpreter of the runes involves time and the accumulation of insight, but this brief introduction is designed to point out how culture was evolving around a model that included both clan and cattle. Both are significantly interwoven throughout the symbolism and interpretation of the runic system. The runes were never connected with astrology, numerology, the tarot, or any other system. They must be viewed in the context of a northern European heritage: the people, their values, and their gods.

4

Woman's Best Friend

Wealth is a barn full of hay, a few hundred pounds of grain, and a cow giving down her milk while cats weave restlessly, ready to spring to the dish by the door.
—Dirk van Loon, *The Family Cow*[1]

A CULTURAL LINK between women and cattle seems an unlikely combination to us today. Most women never see live cattle and have little interest in them. Cows have a presence today as decorative symbols, from the large fiberglass art-cow statues that recently decorated the streets of Chicago and New York, to their widespread presence in department stores, where their whimsical countenances appear on a myriad of kitchen accessories. There has been a surge of interest in all things bovine by giftware manufacturers, who market a profusion of cookie jars, aprons, refrigerator magnets, and the like, all depicting clever or cute cows. At first encounter we may think this bovine decor silly and contrived, yet it recalls a sacredness that began in early cattle cultures where bovines were reared for both milk and meat. The attraction for bovine-related home decor has been around for millennia, since the ancient Egyptians made pottery and bronze items with horned cows. By tacking up a cow calendar, or filling a cute cow-shaped cookie jar, women are unknowingly making a connection to their ancestral past.

Today's ubiquitous black-and-white cattle images represent Holsteins, a dairy breed, and the animals are all cows: females. No Texas longhorns or chunky Black Angus—animals raised for their meat. Women are attracted to dairy animals that signify the female, the domestic, the mother of all. The roots of such attraction are not new; they came about long before Wal-Mart's kitchen design crew decided that anything with a Holstein on it would appeal to female shoppers. Ages ago, women linked themselves inextricably and symbolically with cows.

Cows were important, powerful, and sacred, and images of them have adorned homes, temples, or cooking areas since antiquity. Images of cows and horns decorate many ancient household items because cattle were deities. In ancient Egypt, Hathor was the great mother, the goddess cow, whose body was the heavens and whose udder spewed out the Milky Way. Every day she gave birth to the sun, the Golden Calf. So entranced were Mediterranean cultures with the power of the cow that the name *Italy* meant "calf-land." Milk, *latte* in Latin, was revered and respected for its power to nurture. Goddesses were adorned with headdresses representing cow's horns.

Pre-Christian European creation myths usually involve a cow creatress, under various names depending upon the culture. Myths from the Near East, Japan, and India tell of a world created by the curdling action of milk, and of a universe that was "curdled" into being. Some tell of human bodies being curdled from the goddess's milk. Renenet, a woman in Egyptian lore, held out her inexhaustible breasts to nurture the world, her head adorned with a cow's head or horns. Rennet, the enzyme found in bovine stomachs that causes milk to curdle, was also sacred.[2] The Romans enjoyed a pantheon of gods and goddesses, and one, Cornucopia, is still present today. The cornucopia, or Horn of Plenty, symbolized a cow's horn spilling out the fruits of the earth. The cow as wet nurse to humanity has a long history. But cows are invoked as symbols of milk and the feminine; meat is another matter altogether.

Meat-eating is all about the politics of power as well as nutrition. When a valued item that is also essential can be controlled by one sex, the other will fall into a relationship based on dominance and submission. Males are hunters; women are not. While women in early societies gathered most of the plant foods, they also contributed large amounts of small animal protein. Women used trapping, snares, and nets to catch small mammals. Because children were more valuable than meat, women did not take them along on hunts for large mammals, a situation that was often dangerous. Large mammal hunting techniques also demanded that the hunt be planned in advance, with various hunters given specific positions and responsibilities—which led to the development of politics and social organization within the group. Healthy nonpregnant women would have nursing children, and taking them into the field could not only ruin the hunt (by alerting game with the children's crying) but endanger the child, who was too small to move out of the way quickly. So men dominated the hunt and its organization, and that put them in a position to dominate the rest of society. A hierarchy of meat meant a hierarchy of males. At least in hunting societies.

The anthropologist Craig Stanford argues that despite this situation, women were still powerful, but only to the extent they could manipulate men for better distribution of meat to the women and their offspring. This practice put females at a distance from one another; forming alliances with other females was *not* in their best interest. Other females were simply competition for the best male hunters and their valuable bounty. While males formed strong alliances with others in order to succeed at hunting, females in hunting societies became rivals with one another.[3]

In *The Sexual Politics of Meat,* Carol Adams writes that women in many societies eat a plant-based diet while men eat mostly meat. This second-class diet status for women is common in patriarchal societies. Adams points out that meat has a "gendering influence," in that rules for the division of meat are based on gender. Fat, the tastiest and most energy-laden portion, is usually denied the women and reserved particularly for the men. In many cultures meat is valued because it provides energy and strength for physical exertion; in that case it makes sense that men get enough of it. But it continues to symbolize masculinity, and the idea of a vegetarian weightlifter or wrestler is hard to imagine. Both Adams and Stanford describe meat as an almost sinister product, which led to patriarchies. And indeed it did in prehistoric hunting societies.

The rise of pastoral animal herding societies changed the gendered roles that suited hunting societies. The role of meat—and dairy products—in cattle-based cultures that existed during the long period between the prehistoric and the present contributed to a far different society. Females in herding societies had more equality and more power. The entire social system was built on nurturing skills and attitudes. Females in cattle clans *did* generate alliances, and those alliances were not always all female- or male-dominated; cattle clans were much more gender neutral.

Cattle keepers must nurture their stock, emphasize the breeding and rearing of young animals, and cooperate as a clan to care for the animals as well as process the meat and dairy products. Their commitment is future oriented. In cattle-keeping pastoral societies, culture shapes religion and is itself shaped by ecology. Not only do cattle cultures build a sacred ethic around cattle herding, but they bring the sacred into cattle herding. Cattle are raised to please the deities, and the deities have some role in how successful the herd becomes.[4] Deities were usually female-dominated, with earth mother and earth goddess belief systems. Males, in attempts to re-create earlier domination schemes held in hunting societies, developed the warrior class, whose job was to raid the neighbors and steal their cattle. This allowed men to re-create male alliances and attempt to gain additional power lost as women took over the cattle-rearing. Warriors were indeed essential to protect the herd. New socio-dynamics were worked out, giving males dominance in one area, women in another. But it was clearly a situation in which the sexes depended upon each other for success and survival.

Cattle had to be nurtured, tamed, cared for—a job that women gravitated to because the animals could be kept near the house and fed or watered with the help of children and the elderly. Cows were valued because they were female and could provide additional stock by bearing a calf every year. Milking cows provided extensive dietary protein; and milking was an essentially feminine task, along with processing the dairy products that were vital to survival. As clans began to base their politics and survival on the nurturing of cattle, women came to be seen as more powerful; in fact, the rise of earth-goddess societies coincides with the move onto the grasslands and the adoption of a pastoral way of life.

Walter Brenneman has examined the phenomenon of female deities in cattle culture myth and religion. Writing in the journal *History of Religions,* he moves beyond the cattle-raiding warrior concept of cattle cultures and explains how heavily and frequently their socio-religious mythology invoked or emphasized female deities. Brenneman points out that the Celts, particularly the Irish Celts, had a very positive attitude toward women. Their cultural ideology was earth-centered, their myths focused upon the goddess. Brenneman argues that the Celts of Ireland, Scotland, and the Isle of Man differed greatly from the Indo-European pastoralists who invaded India, Greece, and Iran. The difference was due to the ecological contexts of the two peoples.[5] Irish and Scottish Celtic culture was centered on the cultivation of cereal grains at a subsistence level; it never achieved the surpluses or the population densities attained by the Near East agriculturists, which led to specialization and urbanization there. Instead the Celts of northern Europe lived in small villages; the homes of clan chieftains were seldom significantly larger than the rest of the group. By the fifth century B.C. these Celts had blended cultural elements from southern Russia and the Iberian Peninsula to become a warrior aristocracy based on animal power: horses and cattle. For them, animal power did not reside in the male-dominated world but was attributed to the earth, which was viewed as a female goddess. The earth goddess was seen as the source of power in the pastoral nomadic societies of northern Europe. Male leadership was in the forefront of society, but it depended upon female deities.

The Celts valued agriculture, particularly cattle and horses.

They buried animals with their owners, and their religious rituals centered on animals and the goddess figure.[6] Because the goddess was central to survival and religion, the Celtic cattle culture is empowered by the feminine; it is not the patriarchy found in such nomadic cultures as India, Greece, and Iran. This religious symbolism extends to culture, too, and shapes the roles of women and men in their respective cultures. In the Indo-European world, kingship derived from a sky-god; for the Celts, world power flows from mother earth.

In some Celtic myths the cattle belong to a woman, symbolizing mother earth. When the cows are stolen, it symbolizes the end of life. When they are recovered, rebirth occurs.[7] Other scholars have compared cattle-raiding to bride-stealing: both are equated with fertility, and both women and cattle are highly valued in such societies. In many of those myths, a woman may take the place of the cow in the myth without changing the basic story—cow and woman, or cow and goddess, are one. In the Irish tales, the goddess is identified with the earth and plays the role of mother, lover, and warrior. She is the giver of all life—and thus cattle—and is identified with cattle and the earth.

The Celts were warrior-farmers. They brought ironwork to Europe and are known for their weaponry, but more significantly they brought iron tools: axes, plowshares, and scythes—tools that made growing and harvesting fodder for cattle much easier. They practiced intensive farming and cattle-raising and dominated most of Europe for more than four hundred years before Rome spread its empire outward, replacing many Celtic practices with Roman ways. Culturally, Celtic art and design are the underpinnings of European traditions, but they were pushed to the perimeters of the British Isles, mainly Wales, Cornwall, Scotland, and Ireland.

Celtic religion emphasized May Day—Beltaine—the day marking the divide between winter and summer. To purify the cattle, the herd was run through sacred bonfires to sanctify them for the coming summer grazing season. Celtic feasts, held in the tribal leader's home, included boiled pork, beef, ox, game, and fish, along with honey, butter, cheese, curds and milk, wine, mead, and beer.[8] Celts were the first to use soap, made from beef fat and wood ashes. Celtic currency came in two forms: cattle and slaves. One female slave equaled the value of six heifers or three milk cows. Females, not male slaves or male cattle, were valued more highly.[9]

Children of freemen (typically those who held seven cows and a bull, seven pigs and a sow, seven sheep and a horse, and enough grazing land to feed seven cows for a year) were sent to live with other foster families. A cultural tradition that lasted through the Middle Ages in Britain, the foster family was paid for keeping and training one's child. Boys cost six heifers or one and a half milk cows; girls' fees were set at eight heifers or two milk cows. Girls cost more to foster out because boys were thought to be less trouble to raise.[10]

Around 350 B.C., fifty years after entering Britain, the Celts entered Ireland. They probably came by sea, from the Iberian Celt group, and became the Irish. (Ireland is the only Celtic nation-state in today's world—all other Celts were absorbed into other cultural groups.) Between then and 500 A.D., there was no written Irish history. Thomas Cahill, writing in *How the Irish Saved Civilization,* describes these early Irish as "an illiterate, aristocratic, semi-nomadic, Iron Age warrior culture, its wealth based on animal husbandry and slavery."[11] Ancient Ireland had hundreds of clan chiefs, each of whom ruled a few dozen cattle-ranching families. "Rustlers" more aptly describes them, according to Cahill. There was no law—simply a cycle of cattle raids directed by one extended family against the neighbors.[12]

Cahill describes the women of the Irish tales as "all women who, in life and death, exhibit the power of their will and the strength of their passion."[13] The Irish-Celtic goddess Brigit was worshiped throughout Ireland. She watched over women in childbirth as well as nursing ewes and cattle, and was a healer. She was a patron of crafts and poetry. When Christianity arrived in Ireland, Brigit underwent a change to St. Brigit, the most important female saint in Ireland. She was said to own cows that gave a lake of milk, which could provide a never-ending supply of food for the poor.[14]

Ancient Celts sacrificed cattle too, but they were the more expendable bulls and oxen. The druid responsible for crowning the high king of Ireland first drank blood from a sacrificial ox, then spent the night sleeping inside the animal's hide. The ox blood and hide would pass on dreams revealing whether the wrong person had been chosen king.[15]

Celtic myth and history recorded by the Romans, who lived at the same time, reinforces the idea of Celtic women as strong individuals. Many mythological figures, such as the Hag, sort of an "earth mother," and St. Brigit, as well as Queen Maeve, all owned and were caretakers of their own cattle herds. Cattle-keeping was not a male-dominated, economic base; women raised their own. Cattle could be kept easily by women: heifers were kept for breeding while young bulls were butchered before they grew too big to handle. The environment made it easy for women to build stone fences, which were tedious and took stamina but not great physical strength. Grass hay could be cut with hand scythes, then bundled and stored for winter. Butter and cheese, made by hand at home, were kept cool stored in pits dug in the ground—ancient root cellars. The entire process could be handled by females, with animals sent out to graze under the watchful eye of children. While men were away on long voyages (such as the Vikings) or at war (like the early Irish), the women tended the cattle for years while the men were absent. Cattle-keeping lends itself to a matriarchy, also evident in African cattle-keeping societies.

Women's involvement in cattle-keeping is evident in folklore and myth. The most famous Irish tale describes how the warrior Queen Maeve (or Medb, meaning mead, or "she who intoxicates") sought only the best bull for mating her cows.[16] Called the *Tain Bo Cuailnge,* "The Cattle Raid of Cooley," the story was probably first told about the time of Christ; the written manuscript, in the eighth century, dates from an earlier oral tradition. Ireland remained isolated, so practices and myths held on longer there after they had disappeared or changed elsewhere. The cattle culture described in the *Tain* lasted perhaps until the introduction of Christianity. Scholars argue over interpretation of the *Tain.* Some see it as a conflict between Celtic-Aryan father-dominance versus the pre-

Celtic dominance of mothers in pre-Celtic Britain. Without doubt the major character is not the hero Cuchulain but Queen Maeve, the matriarch ruler of the land. Maeve was a goddess when the tale was first told; not until centuries later did Irish tribes have queens or kings.[17]

As the story goes, Queen Maeve was the real ruler of ancient Ireland, even though her husband, King Ailell, was ostensibly the monarch. She "ordered all things as she wished, and took what husbands she wished, and dismissed them at pleasure; for she was as fierce and strong as a goddess of war, and knew no law but her own wild will."[18] She was tall, with a long pale face and masses of hair as yellow as ripe wheat. She owned her own cattle herd, which was serviced by a prize red bull with a white chest and horns, named Finnbenach. One day Maeve and Ailell were arguing who owned the best herd of cattle, and Ailell teased Maeve because Finnbenach had moved in with Ailell's herd of cows. Ridiculing Maeve for not being able to control the bull, Ailell taunted her for not being able to manage her own cattle. Furious, Maeve went to her head steward and asked if there were a better bull anywhere. He told her about a Brown Bull of Quelgny, "the mightiest beast in Ireland." So Maeve put together a plan to get the bull for her herd. The Brown Bull was in Ulster, which was out of her domain, and the Ulstermen would not part with it even when she offered to rent it for a year and return it. Maeve planned an attack on Ulster and enlisted her husband to help her. She was eager "for fighting, for glory, and for the bull, and Ailell, to satisfy Maeve."[19] Clearly, Maeve was willing and able to fight for her right to operate her own cattle herd as she chose.

The hero is Cuchulain, a young man who defends the Brown Bull and is said to represent the sun god of Aryan mythology. He loses to Maeve's men, and the Bull is captured. Once the Brown Bull is Maeve's, however, all is not well. The Brown Bull meets up with Finnbenach in Ailell's pasture, the two fight, and the Brown Bull kills the other. Pieces of bull flesh are tossed all over the land; the prize Brown Bull, exhausted, begins bellowing and vomits black gore. He falls down dead. Maeve and Ulster draw a peace. The cattle raid is over.

Maeve, the warrior queen, reigned in Connacht for eighty-eight years. She was described as "the kind of woman the Gaelic bards delighted to portray." Not gentle and modest girls, they possessed a "fierce overflowing with life." Thomas Cahill, writing of the cultural achievements of the Christian monastics whose arrival changed ancient Ireland, says that "it would be reckless overstatement to claim that women possessed equality in Irish society."[20] Things clearly changed with the arrival of male Christian missionaries in Ireland, and women were no longer as powerful and independent as they had once been. Mythic Ireland, as with most of Celtic society, had been based on an equality between the sexes.

THE VIKINGS were another northern European cattle culture in which women were active in cattle-keeping. Although we think of the Vikings as part of a maritime and seacoast society, in fact they were farmers and stock raisers. They lived near the sea, but their way of life centered on the keeping of livestock, and cattle were their mainstay. They cultivated grains and vegetables, but the harsh northern climate made livestock more reliable. Norse stock raisers kept their cattle and horses in stables during the long winter and fed them from stores of fodder harvested during the summer. Vikings ate mostly beef, some mutton, and plenty of dairy products. Sheep were raised for their wool, valuable for Norse textiles, and were allowed to pasture outdoors year round.

When the Vikings expanded their domain between 800 and 1050 A.D., they pushed out into the rest of Europe, going "a'viking" to return with plunder from abroad. They spread out in colonial settlements in North America and Greenland, and traveled far into the heart of Europe and the Baltic to plunder. Their expansion was possible only because they kept livestock—and took many along with them. Viking cattle were small by today's breed standards, which made them easy keepers, particularly when they were crowded into the Viking longboats along with the family.

The Norse settlements in Iceland are fascinating because they reveal so many interesting dichotomies. The Norse settlers were remarkably healthy; their remains show an absence of nutritional deficiencies and much better general health than groups on the Continent. They ate almost entirely meat and dairy products. Their lives were based on livestock keeping, hunting, and fishing. There are plenty of cattle and sheep bones for archaeologists to sift through, revealing that the cattle bone remains grew larger as the settlement matured. At the beginning of the Icelandic settlements, people were frugal with butchering cattle, but as the economy matured and the herds grew, they were able to rely on beef as a dietary mainstay.

Obtaining all-important fodder for their animals structured the basic economy of the Viking settlements. Farmsteads were spaced just far enough apart to allow for obtaining hay from surrounding pastures. The largest and richest farmsteads were located in grassy meadows that could be cut to provide stocks of hay for winter feeding. Their locations were selected to provide hay, not access to the sea or river travel. Meadows that flooded each spring could provide ample grass for several cuttings (using a hand scythe) of grass hay each season.[21] It probably took at least a decade for the number of cattle to grow so that newcomers did not have to continue bringing their stock with them but could purchase cattle from earlier arrivals. The Vikings, with their emphasis on cattle, were like other cattle cultures in that women were much the equals of men, with much more standing and freedom than in crop-growing patriarchal societies.

The historical tension between patriarchy and egalitarianism is essentially the friction between agrarian and pastoral lifestyles. Patriarchy flourished within agrarian societies where ownership of land and control of labor (and people) was held by males. Agrarian societies, with their complex class system, government, and occupational specialization, were characterized by clearly stratified gender roles. As the historian and clinical psychologist Catherine Manton notes, "The subordination of women actually has been most intense in agrarian societies."[22] That subordination was reinforced with feudalism, colonialism, and ultimately the Industrial Revolution. In spite of so many obstacles, women from some agrarian cultures were able to maintain a tenuous hold on their bond to cattle.

American settlers are one example. On small farms, where most Americans lived in the mid-1800s, women relied on a cow or two for economic stability. Making and selling butter had been a woman's route to financial freedom for centuries. During the era of American expansion and industrialization, enterprising women made and sold butter to provision sailing ships and for a growing urban market. By the mid-nineteenth century, making and selling butter had replaced the home spinning and weaving industry of colonial times. Joan Jensen explains that women on Western farms continued the pattern they had learned in the East of producing butter for urban markets. "By 1860 eastern farm families had already come to depend on butter for a cash income to supplement income from grain surplus . . . making butter now became the chief occupation of farm women and girls. . . . As the frontier became settled in the West, more women turned to butter making as a way of supplementing the farm income. By 1910, women on the plains of Montana were using butter money to buy windmills necessary for survival of farms in the waterless land."[23] Into the twentieth century in Wyoming, women like Elinor Stewart kept ten cows, which could provide enough butter to buy flour and gasoline, essentials for survival in a modern farm economy.

Butter was ideal for women's entrepreneurial energies: it would keep better than fresh milk because it was salted, and it was easier to transport to market than milk because it was compact and solid. Cheese took more work and longer for aging. It needed elaborate preparations and greater temperature control. Country stores took butter in trade, allowing women to barter for items they needed. Women sold garden produce, eggs, and poultry too, but butter was the economic mainstay.[24] It was a cultural commodity that had been brought from northern Europe, but even the Osage Indians of Oklahoma turned to it as a cash source, producing thousands of pounds of butter for the market each year.[25]

Today women in the United States no longer rely on butter for economic freedom—even if they wanted to, they would find that the market has been taken over by industrial giants and threatened by margarine. But over most of history, a milk cow has represented economic freedom for a woman. An excerpt from a woman's memoir of life in the 1930s in the drought-ridden hills of southern California reveals that poignant relationship which once was so all-important to women. From Joan Jensen's feminist history, *With These Hands: Women Working on the Land,* here is a portion of Judy Van der Veer's experience one summer:

> "I've most got Wilbur talked into the notion of lettin' me buy a cow," said Amelia hopefully to me one day.
>
> She was a huge drab woman and the only time I ever saw her big honest face light up was when she talked of cows. Often she came in my corral at milking time to buy a jug of warm milk. She always stayed to admire the cows, the fat heifers and calves. She was like a cow herself, a great full-bellied woman, placid and wise. Plainly she remembered the lush meadows of her father's farm "back East," she could shut her eyes and see the string of spotted cattle coming home at milking time.
>
> And now here she was in this arid land where one had to pour gallons of water on the earth and loosen it with a hoe so it wouldn't dry like cement, where one felt guilty of robbing vegetables of water in order to keep a few flowers bright around the dooryard. On their land she and Wilbur had vegetable gardens and fruit trees and a little vineyard; their live stock consisted of pigs and poultry. Amelia would be satisfied if she could only have a cow. If you had a good cow there would be milk for the turkeys and ducks and cream for butter, and maybe enough to raise a calf that would grow into money.
>
> "Wilbur thinks it would cost too much to feed a cow. But in summer I could raise a little patch of field corn and sudan grass and beets; and in winter there's green grass a-plenty around our place. Why I'd work my fingers to the bone to feed a cow. She'd pay for herself in no time, anyway."
>
> Amelia would watch my slow milking with disdain. "Let me

pump that cow awhile," she'd exclaim, and I'd relinquish stool and pail to watch her beautiful mannish hands stroke thick streams of milk from William's udder.

The cows and calves loved her for her quiet understanding. Even Fawn, William's high-bred nervous heifer would nose around her affectionately, and Amelia would run her hands over each one, gloatingly, telling me of their fine points.

"Look at that there nice long tail and straight back. Did you ever see anything as big as this heifer's eyes! And look what long tits she's got. She'll make a fine cow for sure. Wish't she was mine."

I wished so, too. But I knew that heifer was worth far more than Amelia could afford to pay. How I longed to be rich enough to give it to her or sell it to her cheaply.

One day when I was riding down by the river I saw old Henery riding along on his mule, "bunching up" his little herd of cattle. Henery's cows never do well. They look gaunt and moth-eaten, and his calves are always pot-bellied and lousy. He can never find enough good pasture for them and he hasn't any money with which to buy hay.

"Got a heifer I orta sell," shouted Henery, "she lost her calf and I ain't got another one to put on her. Might let her go cheap if t'was cash. I need a few dollars."

I thought about Amelia. For a long time she had been saving egg money and doing without things for herself.

Her face was bright when I hurried home and told her about the bargain.

"Don't get your hopes too high," I cautioned. "I didn't see the heifer, she's probably an awful scrub, and maybe your husband won't think she's worth buying."

"Well, it's my money I've saved," said Amelia. "I've sewed for the neighbors till I most put my eyes out; I've made underwear for myself and the kids out of feed sacks; I've chased the hens around after eggs; and if that heifer looks at all good to me, I'm a-gonna buy her. If Wilbur don't like it he can jest go take a jump somewheres."

A few days later when I rode past Amelia's place she came out and hailed me. "Come in and see Rosie," she cried, "I bought her jest like I said I was gonna do!"

When I looked at Rosie I saw a stunted, dull-eyed, scrawny, red heifer. She looked to be half Hereford and half Jersey, and I wondered if she would give as much milk as a goat. Rosie was eating cornstalks as if she had never tasted anything so delicious in all her two years, and probably she hadn't.

But while Amelia petted the heifer's rough coat and talked about her I began to see what Amelia was seeing—the Rosie of the future. Red coat shining from much brushing, sides sticking out with fatness and pregnancy, eyes clear with good health, bag swollen with milk, curved horns and neat hoofs polished, and a beautiful red curl on the end of Rosie's tail. Suddenly I saw it all just as Amelia did, and together Amelia and I waxed poetical over that gaunt red heifer.

And would you believe it, in a few months Rosie began to resemble the picture of her future self! Amelia lavished care on that little cow and Rosie responded by growing fatter and "coming upon her milk." Her eyes were bright now, and they burned with affection when they were turned toward Amelia. She would stretch out her tongue to lick at Amelia's dress or arm, she mooed with joy whenever she saw her. She was bred again, too. Amelia had saved up five dollars to pay the breeding fee on a fine Jersey bull.

"Yes, sir," said Amelia, "jest wait until she comes fresh. She'll give as much good milk as any cow in this here valley. Feed a cow good, and take care of her right and dry her up at least two months before it's time for the calf, and she'll give a lot of milk. And when she's had a couple of more calves and is in her prime, some dairy man will be just achin' to buy her. But I won't sell, no sir-ee. I'm sacrificin' to make a fine cow out of her and I'm a-gonna keep her for myself."[26]

WOMEN AND COWS share another, more recently discovered bond: hormones. Cows are nature's most protective mothers—they will not hesitate to attack anything that threatens their calf. Scientists have found that the cow's pituitary gland (located next to the brain) contains a powerful hormone that drives maternal behavior. The hormone has been extracted from cow brains at slaughter, then administered therapeutically to pregnant women as *oxytocin.* Given intravenously, it causes pregnant women to go into labor, saving the lives of both women and infants. By the 1970s oxytocin was commonly used to put women into labor who otherwise would have been forced to have cesarean surgery or not been able to give birth at all. As Dr. J. H. M. Pinkerton declared at an oxytocin research symposium in 1965, "When the history of this particular era comes to be written, one of the great advances will be seen to be that obstetricians were no longer prepared to allow the foetus to die *in utero,* content with the knowledge that they had not killed it with their own hands. In other words, we have come to realize that the uterus is a very dangerous place to be in certain circumstances and that the baby must be got out of there somehow or other."[27]

The thousands of women today who raise cattle, and find themselves anxiously waiting up nights during spring calving time, share a maternal bond with their animals. They nurse, and coax, and pull the calves from the mothers if needed to save the calf or the cow. Linda Hasselstrom, an environmental writer and Wyoming rancher, calls those cattlewomen "midnight heifer midwives."[28] Nancy Curtis, editor and publisher of her own *High Plains Press,* writes and ranches in Wyoming. She tells about a call from a New York editor that caught her during calving season. She asked her mother to take the call, instructing her, "Don't say I'm out checking on my first-calf heifers. Say I'm meeting with my production staff." Her staff, she reported, turned out some nice calves that year.[29]

In her essay "Cowmoms," Nancy Curtis tells about the motherhood-related links between women and their cattle. She writes:

> I've seen a cow disoriented for a month after losing a calf, returning to the site of the birth, searching, calling. It tugs at me in a deep spot where the mothering instinct is never completely buried.
>
> One winter an old redneck cow calved prematurely and lost her

calf. I found the remains by following her tracks to a gully. Blood in the snow and a few fragments of soft bone were all that was left. The coyotes had found the calf first. . . . [The old cow] disappeared for days at a time. She didn't come for food or water, surviving by eating snow and grazing on dry patches of last summer's grass. Then I'd spot her—slow moving, skinny, losing hair, dull-eyed. I told myself if I saw her alive again, I'd take her to the vet.

We did tests. "Nothing wrong," the vet said.

"Maybe she's depressed," I ventured. I didn't have the guts to suggest that maybe she was traumatized. Maybe she'd watched the coyotes eat her calf—maybe it was already dead, maybe not. The vet didn't laugh at the idea of depression.

"What should I do with her?" I asked. "Is she going to be all right?"

He said, "Sometimes they never get over it. If she was mine, I'd send her to the canners."[30]

5

Cattle Culture Comes to the Americas

> Before him lies a boundless continent,
> and he urges forward as if time pressed
> and he was afraid of finding
> no room for his exertions.
> —Alexis de Tocqueville,
> *Democracy in America*[1]

WHEN WE THINK OF North American cattle ranching, we think of the era of open-range cattle grazing on the Great Plains, which holds a mythlike place in our national memory. But that era of the huge herds, wide-open spaces, and trail drives lasted a mere twenty years, from the end of the Civil War to the mid-1880s, when weather and overstocking resulted in the collapse of the cattle market. We are romantically tied to that short-lived Western vision, yet where did it come from? How did Western cattle ranching erupt seemingly out of nowhere to capture our hearts and imagination?

Cattle ranching has thrived in a wide variety of environments: tropics, pine barrens, prairies, lowland plains, mountain ranges, and meadow wetlands. Cattle are adaptable creatures, which do well in nearly all conditions in temperate climates. Cattle ranching was never a business that originated in harsh environments, rather it was pushed out of more favorable locations and allowed to exist beyond the fringes, in a refuge of its own. Cattle herding worked best at the fringes of settlement; where new pastures could be easily moved onto, and where land was cheap. Cattle ranching was a mobile business, one that moved to a fresh habitat whenever overgrazing wreaked havoc with grazing lands and commercial markets pushed herding to the far corners. Land close to the market commands higher prices because it costs less to transport products to market. Thus intensive agriculture, such as tilled field crops, dairying, and feedlot operations, locate closer in to profit from market proximity. The hinterlands become cattle ranching districts because grazing is the least intensive form of commercial agriculture. As population grows and disperses, grazing gets pushed farther out onto the least productive lands.[2]

While Americans have viewed cattle ranching as uniquely Western, University of Texas professor Terry Jordan calls that image "largely illusion and myth." "Ranching was not a product of the frontier or the semi-arid West," he argues.[3] Ranching as we know it did not originate in the American West, Latin America, or even the East. It was a cultural activity that immigrants carried out of the eastern Mediterranean and the Nile as they swept out across Europe and Asia. As settlements grew they were pushed farther out, eventually relegated to the isolated fringes of Scotland, Wales, Ireland, Brittany, and the edges of the Iberian Peninsula. By 1500 A.D. the cattle-culture belt extended from Scandinavia along the Atlantic coast of Europe and into Africa. Even then the cattle folk had been pushed to the periphery of civilization's centers. The discovery of the New World provided their next move—the ideal solution to their situation. The cattle cultures of the Old World had been pushed nearly to oblivion when a whole new hemisphere beckoned. "At almost the last moment, a reprieve had come," Jordan explains.[4]

Three major centers of cattle herding contributed to American ranching: the southwestern Iberian Peninsula, the British highlands, and West Africa. Spaniards and Portuguese from Iberian ranches were the earliest to graze cattle in the New World. The Portuguese had a unique colonizing system: they shipped cattle to areas they planned to settle later. The stock increased naturally by the time colonists arrived. In the sixteenth century they used the technique in the Azores, the Cape Verde Islands, and Madeira. When the Spanish tried it later on the European mainland, it was not as successful. Iberian cattle raisers had no interest in a dairy industry; they sought wide-open grazing lands and quickly adapted to Latin American grasslands. The British highlands cattle culture, in Scotland, Wales, and Ireland, was the foundation for many ranching practices—open range, overland drives, pasture burning, and Western livestock law. The term *cowboy* was used in the British Isles long before colonists embarked for the New World. Cowboys or herders were low-status, often indentured servants who worked for a wealthier chieftain or lord. Cattle tenders were male or female, young or elderly, and they herded cattle on foot, using rocks, whips, and trained dogs to push the cattle along.

The theory that Western cattle ranching was of Hispanic origin has been popular, and Jordan notes that the idea of ranching moving into the arid West because of the environment, combined with Hispanic origins, is "almost irresistible." While Texas was a pivotal region in Western ranching, though, not all ranching in the American West derived from the Texas model. "In ethnic terms, Western ranching reflects a unique mixture of groups, a blending of British, American, Hispanic, and probably also French, German, and Amerindian influences," Jordan writes. "The crucial early mixing occurred in one confined locality, lowland South Carolina, the southern fringe of the English colonial empire in North America, where a favorable juxtaposition of Britons and West Africans occurred."[5]

The influence from West Africa came with slaves who were brought to the Carolinas and other Southern colonies, which became the early center of the open-range cattle industry. After 1670, slaves from Gambia arrived with previous cattle experience. Occasionally traders brought both slaves and cattle from Africa to the colonies. While the cattle cultures of Africa were different from those of the British Isles in that they were more like pastoral nomads than ranchers, they brought cattle handling skills that spread into American ranching practice.[6]

The American Western ranching culture that emerged from these three influences—Hispanic, British, and African—was a bit of each but largely a Celtic, or British, highland system. The cattle culture of the Celts emerged as the practice best suited for the West, and though it was shaped by ideas from Texans and Californians, too, the Celtic system is the one that survived. Still, it would be wrong to describe the North American cattle ranching system as one particular practice; it was an amalgam of practices and ideas, suited to specific environmental conditions and shaped by politics. As lands shifted to Spanish, French, British, or American rule, the cattle industry changed to suit the times.

South Carolina was the "hearth area" or source for large-scale, Anglo-American cattle herding in the colonial era. By the mid-1700s the colony had almost 100,000 cattle, and annual slaughter was around 12,000. Beef was the major export, barreled and shipped to the West Indies slave plantations.[7] No other colonies had the combination of climate, grass, link to the West Indies market, British-African cultural heritage of cattle-raising, and Gambian slaves with cattle-tending skills. By the time of the American Revolution, Carolina's dominance had faded due to overgrazing, cattle disease, and cotton planting as well as greater numbers of settlers who diminished the open-range pasture. As the cattle culture spread west it picked up French and Spanish techniques, but early on the Carolina cattle culture was the industry's foundation.

Carolina ranchers each held seven hundred to a thousand head, running the animals on the open range, branding, and using roundups. Animals were allowed to range freely and were rounded up from time to time to send to market. People and dogs pushed the herd together and drove them into a rail-fence cowpen. Lacking horseback roping techniques and saddles with horns and double cinches (which would come later from Spanish-American influences in Texas), Carolina herders used salt, whips, and trained dogs to control the herds. Salt was effective where there were no fences; it kept nearly wild cattle near the cowpens and accustomed to humans. On Saturdays, herders scattered handfuls of salt in the grass or on flat stones. Once the animals were used to it, herders could withhold salt before an overland drive, then use it to lure the cattle along the route.[8]

British herders had begun using trained herd dogs in the fifteenth century, and they had long been important in droving. Dogs were taught to chase cattle until the bovines formed their classic defense circle, which effectively bunched them together. "Bulldogging," a term from eighteenth-century Britain, refers to the working of cattle and hogs with bulldogs. The custom spread to the colonial South in the 1700s. These herder dogs were called Catahoula, Tennessee brindle, or leopard dogs. They were medium-sized hounds, spotted or striped in random patterns, with light-colored eyes. They could bring down a wild cow or hog by grabbing the animal's nose, lip, or ear and pulling it to the ground. The best cow dogs were trained to pull or "cut" a particular cow from the herd. Dogs were used more often than horses, but horses were sometimes taught the canine techniques. The quarter horse, bred for speed in short bursts, was trained to operate as dogs did. Using back-and-forth movements, it moved a cow away from the herd just as a dog would. A rider on a trained cutting horse is unnecessary because, in Jordan's words, "a quarter horse trained as a cutter is well named—it is three-quarters dog." Modern rodeo events still echo the early practice of working animals to the ground by hand. "Bulldogging" was an innovation of Bill Pickett, a black Texas cowboy of South Carolina heritage. Pickett used his teeth like cowdogs did, to bite the upper lip of the animal in order to bring it down.[9]

Bullwhips were standard in the American Southeast; they were about twenty feet long and made of rawhide strips braided together with ends left loose. In the hands of a skilled drover they were highly effective in maneuvering cattle. Some sources claim that Georgia-Florida "crackers" were named after the sound of the whip as it snapped above the heads of their oxen, but that's not true. It was an English term from the 1400s, used to refer to someone disparagingly, or as a liar.[10]

About 1750 a distemper hit South Carolina, wiping out many of the cattle, and back-country bandits stole cattle and tore up cowpens. Disorder and anarchy was common, and the cattle industry in the region never recovered. By 1840 most of the South Carolina cattle herders had moved to the pine country of east Texas, or into Georgia and Florida. In the early 1700s, Cherokee Indians began herds of their own, so did Seminoles, Creeks, and Chickasaws. Later the Indians took the cattle culture to eastern Oklahoma when they were forced to move to Indian Territory. It was during this move into Spanish territory that the industry picked up elements of French and Spanish practice. White and Indian herders had to alter their techniques in the new environment; open country was hard to manage with dogs, range grass had a high saline content that made salt less valuable, and the use of horses and ropes replaced dogs, whips, and salt.

THE HERITAGE AND PRACTICES of Scotland, a land steeped in cattle-keeping, have been particularly influential in the American West. In 1970, John McPhee described a trip he made to his ancestral homeland in the Hebrides Islands, twenty-five miles west of Scotland. There he found a community of seventeen crofts (small cattle-based farms) and seven farms (a *farm* being more than forty-nine acres under cultivation). The land was owned by a laird, or landowner, who held it through inheritance and rented the parcels to tenant crofters. The small-farm society of Scotland had been primitive, each family tilling a small patch of ground and cattle grazing together on a commonly held pasturage. Families lived in houses built of stone with thatch roofs. Peat blocks were gathered locally and burned as fuel. "Cattle and horses lived in the houses, too, or in adjacent byres [cow barns]," McPhee pointed out. "The animals often used the same entrance the people used."[11] It is hard to imagine people who lived closer to their cattle or their clan.

Clans—small family groupings—had formed regional governments in Scotland for seven hundred years until the Battle of Culloden Moor. English armies defeated the Scottish clans at Culloden in 1745, and afterward the clan chiefs became the landowners—lairds—in the modern sense. The English government replaced land held in common with land held in fee simple—with no restrictions on the transfer of ownership. Private ownership was put in the names of the chiefs, their clansmen were made their tenants, and the system of small-scale cattle ranching that had worked for seven hundred years was ended, replaced by the English legal system. Eventually the chiefs "sold them out," McPhee explains, to absentee owners who held the land. The new owners saw that sheep were more profitable than tenants and cleared the land of crofts in favor of larger pastures. The Scottish Highland cattle raisers were evicted by the landowners in what has become known as the Highland Clearances. People were forced out of their homes, the houses torched, the livestock slaughtered or run off. It was an era of terrorism and what many call genocide; the impoverished, homeless Scots had no place to go. McPhee plaintively notes, "The people leaving sometimes had to drain blood from their cattle and drink it in order to survive."[12] But with the development of the power loom, England's factories required ample supplies of wool. Sheep were brought in by the hundreds of thousands to stock the pastures. The era of cattle and clans was over.

In the United States, the response to the Clearances was negligible. Many felt the Scottish landowners had done the right thing by putting the land into more profitable production. Ironically, Harriet Beecher Stowe, who by that time had made her reputation with *Uncle Tom's Cabin* and its heartbreaking portrayal of Southern slavery, felt no empathy for the Scots. In her *Sunny Memories of Foreign Lands,* she put a positive spin on the landowners' actions. Clearing out the primitive herding families and putting the land to optimal use made sense to many proponents of scientific agriculture and industrialism.

"Second sight," the ability to foresee events, was highly esteemed in the Highlands. A famous seer, Kenneth Mackenzie, is supposed to have foretold events of significance. His ability at second sight was highly regarded. A hundred years before the Culloden battle and the end of the clans' power, he was said to have warned that "the clans will flee their native country before an army of sheep."[13] By mid-1800 it had happened just as Mackenzie had predicted. Entire villages were evicted and shipped by boat to North America.

The cattle-raising people of the British Isles were pushed to America in droves, so to speak. Scots and Irish were shipped to Barbados and mainland colonies in the seventeenth and eighteenth centuries as bonded labor or transported as criminals—some for crimes as innocuous as "stealing bread and cheese."[14] Cheaper than slaves, indentured servants cost half as much and were obligated for four to five years of labor—criminals much longer—and if one died, no financial loss ensued. Scots and Welsh were deemed the best servants; Irish the worst, largely because they were Catholics, which placed their religious beliefs and loyalties in doubt. Maryland and South Carolina placed a duty on imported Catholics, and the Barbados Assembly required Irish servants to carry passes. Buying and selling indentured servants became a brisk business between the British Isles and the colonies. People resorted to selling themselves into servitude, and selling children was commonplace in the northeast of Scotland; the term "kidnapping" derived from the seventeenth-century practice of stealing children and shipping them to the colonies for sale. Scottish newspapers carried advertisements luring potential recruits with the promise that after four short years of service they could go into business for themselves or gain their own land.[15] Scots, with a legacy of clan battles and war with the English, were well suited to frontier fighting, and many had been purchased as indentured servants to man frontier forts in South Carolina and Georgia. Another reason for recruiting Scots was their rural background. Henry Laurens, a landowner in Florida, wrote instructions to his recruiter in 1766 to find Scots who were "simple & unacquainted with the tricks & vices of the Town."[16] Many recruiters hit the country fair circuit in Great Britain, where they found willing and able rural folk who signed up for transport to the colonies.

Women too were frequently indentured servants. One political writer even suggested that female convicts be transported and given in marriage to Indians in order to cement Britain's alliance with the natives of the New World. But most were bought to perform labor, women being put into fieldwork only if they were "nasty, beastly, and not fit" for other duties. "Dissolute or lewd" women were also sent to the colonies in a steady stream.

Regardless of how men and women arrived in North America, they had no other choice but to come. One eighteenth-century writer noted, "He hath no alternative, but to starve, or emigrate."[17] Once they had served their time and were freed—if they indeed survived—servants found it difficult to establish themselves in the colonial economy. Unable to compete as free labor in a system dominated by slavery, many went into the pirate trade, privateering from Newfoundland to Guiana, or they made their way to the edges of the colonial frontier, where they might manage to obtain a few cattle which they ran on open range.

Forrest McDonald and Grady McWhiney, two Southern scholars who have studied the Celtic influence on American development, observe that "The first and most important thing to know about Celts in America is that they tended to settle in different areas from those settled by the English and other Germanic peoples: by and large, the Celts went south, the English north."[18] The pattern of settlement in North America created socio-political and cultural foundations that affect politics and culture even today. The Celts came to America first in the Scotch-Irish migration to the Philadelphia area, from which they headed into the backcountry where land was available. From there they moved into the Carolinas and Georgia. Virtually none went to New England because most ordinary families could not afford to locate there—some communities even required letters of recommendation from aspiring settlers.[19] New England settlers were largely from eastern and southern England, and shared many cultural traditions that were not common to people living in other areas of Britain, Scotland, and Ireland. When New Englanders spread west to settle, they moved across the Northern states, into the upper Midwest, and eventually to Oregon and Washington.

Southerners, mostly Celtic in origin, moved west across the

southern part of the continent, and by 1850 the South was more than three-quarters Celtic.[20] This heavy concentration of Celtic people left an imprint of language, social organization, and the traditional Celtic means of making a living by raising cattle. By the time the Celts made their way to North America, their clan system and Gaelic language had been eroded by British influences and were nearly gone. Their cattle culture, however, was intact.

Cattle-raising had become a significant economic activity in Britain, and droving cattle to market had become well established long before Celts and cattle ventured to North America. Scotland's main product into the eighteenth century was cattle; by 1378 the Scots exported 45,000 hides annually. Cattle were driven from Highland pastures to the London market in herds—as many as 320,000 passed through the market town of Carlisle in one year. In 1665, Ireland was shipping out more than 14,000 beef cattle a year, increasing to 72,200 in 1798 and, by 1818, 87,771. Besides live cattle, Ireland exported nearly 3 million barrels of beef between 1780 and 1800, along with thousands of pounds of tongues and tallow, and 6.5 million skins and hides. The Irish did not plant potatoes or any other crop until the latter part of the eighteenth century, and the Welsh had a pastoral economy until the nineteenth century when they began working as miners. When they moved to North America, they too brought along their cattle culture.

Although we do not think of Southerners as cattle herders, before the Civil War the value of livestock in the South was greater than the value of all its cultivated crops combined.[21] One reason we do not associate the Southern image with cattle tending may be because Southerners seldom tended their animals. The Southern manner of livestock-raising mirrored the traditional Celtic way: animals were never tended but were marked by ear-clipping or branded and allowed to roam for their forage. Until the twentieth century, Southern land law allowed animals to range freely, a practice that continued in those Western states that were settled by Southerners. "Free range" meant that owners of planted crops had to fence *out* the animal; the animal owner had no liability for its depredations. In fact, roaming livestock are still protected in many open-range areas of the West; if a vehicle kills an animal in a roadway, the driver must reimburse the animal's owner for the damages.

Allowing animals to graze freely was widespread in the areas settled by Scots, Irish, Welsh, and Cornish immigrants. Animals roamed the backwoods as well as anyone's landholdings they chose. In the years before the Civil War, between four and five million hogs and two million beef cattle were rounded up off their ranges each autumn and driven to market, the animals walking up to four hundred miles in some cases. The Southern "plain folk" were self-sufficient because this system was based on nonownership of land, a small investment in labor, and a fairly leisurely way of life. Living at a subsistence level, their only "cash crop" was cattle and hogs, and every family had a few to round up and sell. Even at this economic level, every man, woman, and child in the South consumed an estimated 248 grams of animal protein per day, which is five times today's intake. Largely beef and pork, with some wild game, the diet was supplemented with vegetables and fruits that grew nearly wild in the mild climate, and families were able to subsist with very little labor.[22] Years later, when the Southern poor were relegated to a diet based on corn and molasses, pellagra became rampant. Twentieth-century plain folk in the South were far more poorly nourished than their pioneering counterparts.[23]

Like the Celtic backwoods and Southern folk, Yankees brought a different style of agriculture to New England with them. Coming from England, the settlers in the Northern colonies practiced open-field agriculture, a system that relied on plowing, sowing, and reaping in an intensive yet cooperative manner. They had farmed that way in England since medieval times, quite opposite to the manner of the Celtic peoples. New England was orderly, with a community-oriented spirit. In New England settlements, no one lived out of town; all houses were centered in a planned community while the arable land was divided into equal-sized fields. Everyone's animals were herded together on the common pastureland. The town hired a herder, and he came by each house in the morning, picking up the animals to take to graze. In the evening he would return the animals to their owners, where they were penned up for the night. Anyone who let his cattle or hogs run free and damage crops was fined severely. Villages even hired "fence viewers" to make sure everything was in repair. "Good fences make good neighbors," from Robert Frost's poetry, sums up the New England attitude, which was based on centuries-old English practices.

From these early settlement patterns North and South, pioneers who moved west took their cattle culture with them. The grazing of large cattle herds on open range moved from the seventeenth-century Carolinas to Texas. There the cattle grew quickly, with a good climate and few diseases (besides tick fever), awaiting the end of the Civil War and the beginning of the trail drives to Eastern markets. Confederate ancestors had driven scrawny, near-wild cattle on annual trail drives from Scotland and Wales to feedlots in Essex, Kent, and Sussex. Open range and the absence of fences had long been common in Celtic areas of Britain. Cattle herds were large in sixteenth-century Scotland, where individual owners might keep a herd of four hundred to a thousand cattle, selling off about a quarter each year. Irish herds were just as numerous.[24]

English herds were small. A dozen or so animals was an average-size herd; one of the largest yeoman farmers in eastern England in the early seventeenth century owned eighty-seven cattle—a significant number in that region. England was largely a nation of carefully controlled agriculture. Fences, hedgerows, pens, and laws protected cultivated fields from livestock. Animals had to be kept secure. Mixed farming, the tilling of fields, and the keeping of a few livestock became the norm in Britain. It was orderly, with close control of one's land as well as one's animals.[25]

People whose ancestors had practiced the same herding economics for generations followed their patterns in the American West.

THE LEWIS AND CLARK EXPEDITION returned from exploring the Louisiana Territory in 1806 and opened up the West to the U.S. fur trade. Decades later, farmers and livestock herders led the westward movement across the continent, moving first across the Southern frontier. Wealthy, slaveholding planters did not take the lead in settling the continent; rather, it was the Southern plain folk who were the true pioneers of the Southern frontier. They had initially settled in Pennsylvania, then moved into the backcountry and hills southward, where land was cheap and winters were mild. Behind the plain folk were the planters who settled where fertile soil or waterway transportation made cash-crop agriculture profitable. They generally held over two hundred acres and worked more than twenty slaves. Behind them came the tradespeople and others who settled small crossroads towns.[26]

The plain folk lived in dispersed rural settlements, with relatives and extended family spread out across the land to raise small herds of cattle or pigs. Living on isolated, widely separated farmsteads, they allowed their marked and branded cattle to forage on the unfenced range. Split-rail fences were adopted to protect kitchen gardens and the small tilled fields. Houses were log cabins, similar to ancestral homes in Appalachia and, much earlier, in Celtic Britain.

But the Southern plain folk and their Celtic husbandry practices faltered when they reached the Great Plains. Their stockman-farmer-hunter economy no longer worked there. Woodlands farmers settled along streams and thought the prairies less fertile than the woodlands. They were suspicious of treeless lands, as their ancestors had been, and preferred to locate fields in forests, where the tree cover signaled fertility to them.[27] They needed woodlands to use for houses and rail fences, as fuel, and for pasture. Long a symbol in common with cattle cultures, the double-bladed ax was a mainstay of their lifeways too. They easily adapted to the vast woodlands of Texas and the Southern frontier but held back from settling the Midwestern prairies. They avoided the best soils, locating where the soil was marginal for cropland but where woodland pastures provided grazing and forage for cattle. While land was cheap and abundant, labor was not. Most backcountry folk owned no (or few) slaves, and they were hard pressed to clear, fence, and manure fields in order to put them under profitable cultivation. It was cheaper and easier simply to pasture the edges of the forest, clearing it enough for new grazing, or move on to fresh lands. Because much of the backcountry land was unclaimed and unfenced (it belonged to the U.S. government), it was easy to continue Celtic pastoral customs.

The abundance of land made the practice work. New fields had to be cleared to replace old in a ceaseless procession into ungrazed Western lands. A single range cow needed about fifteen acres of pine forest pasture in order to forage one winter. As more settlers came into the backcountry, the amount of grazing land diminished, causing a continual push westward. Families could sell their lands at a profit and move on as long as there was someplace to go. By the 1830s, backcountry and plain-folk descendants had settled the lower Old Northwest and the Old Southwest, and had begun grazing the New Southwest. Indian depredations were few because government had pressured the Eastern tribes into ceding their lands and had moved them west on the Trail of Tears, across the Mississippi River to Indian Territory. In the years before the Civil War, thousands of plain-folk families moved into Arkansas, Missouri, Iowa, and Texas, where large tracts of unclaimed lands promised opportunity. By the 1860s the grazing lands of Illinois, Iowa, Missouri, and northern Texas were filled with range cattle. They were driven to market by the thousands, marching eastward to slaughterhouses. Cotton meanwhile had become the Old South's most valuable crop and the cotton plantation the popular image of Southern life. But log cabins, corn patches, and free-grazing cattle made up the larger, more accurate picture.[28]

At the same time a new type of intensive agriculture was emerging in Maryland and northern Virginia, stimulated by English innovations. Called "alternate husbandry," it involved rotating crops with grasses and legumes that would restore vitality to the soil. Livestock were contained in fenced paddocks and fed stored fodder. This allowed selective breeding and the possibility of improved meat and milk yields, in contrast to the random breeding of scrubby open-range livestock. This sort of agriculture required a sizable investment, though—cash for buying land and for constructing fences, outbuildings, and storage facilities. Seeds, equipment, blooded livestock—these essentials were not within reach of backcountry farmers.

WHILE THE CELTIC INFLUENCE on American cattle ranching is significant, the Celts in America were not the only cattle herders, and certainly not the first. Erik the Red was actually the first to land cattle in the New World, when in 986 he brought a herd to Greenland along with 450 new colonists from Norway. Cattle grazed easily on the thin green carpet of arctic heath, and the colony thrived. Archaeologists estimate that at one point there may have been 3,000 Vikings living there. The houses were built of driftwood, logs, and sod, with walls several feet thick. The climate was too harsh for grain crops, so the settlers probably ate animal products such as milk and cheese, and the fruits of fishing and hunting. They subsisted for two centuries, but life was difficult in an environment alien to Scandinavia. The climate grew colder in the fourteenth century, and an epidemic of Black Death wiped out a third of the people. The colonists' diet changed to 80 percent fish, revealing that their cattle had not flourished. By 1500 the settlements in Greenland had vanished.[29]

Christopher Columbus brought cattle too. Columbus had taken cattle aboard on his first voyage, but they were carried to provide fresh beef for the crew. None made it to the New World alive. On his second voyage, Columbus brought seventeen ships with fifteen hundred adventurers. He also picked up a herd of cattle in the Canary Islands. After a month of searching for a coastal landing, he put in at Hispaniola. Aboard were ten mares, twenty-four stallions, and an unknown number of cattle. That herd is likely to have seeded the first North American cattle. More ships came with more colonists and their cattle. Eventually they spread out into other islands and made it to the mainland. It is not certain who brought the first Spanish cattle to North America—perhaps Hernando Cortés in 1521 when he set out from Cuba to take Mexico from the Aztecs. But Florida historians claim that Ponce De Leon landed a herd of cattle there months before Cortés.[30]

Within a decade of Cortés's arrival, cattle ranches had proliferated on the mainland of Mexico and along the Gulf Coast. As always, the beef market was quickly flooded with an oversupply, and between 1532 and 1538 the price of beef in Mexico City dropped 75 percent. With cattle in abundance, the Spanish colonists spread out over the continent, taking herds of cattle everywhere they went. Coronado, on his quest for gold, took along five hundred cattle to provide fresh meat on the hoof while he and his men trekked over Arizona, New Mexico, Texas, and Kansas. Catholic missionaries settled down and raised herds of animals, relying on them for the financial support of the missions. With native converts (or slaves), the missions were able to establish the beginnings of industrial beef ranching. By 1700 Texas was home to Spanish missions and Spanish cattle. At the time of the American Revolution, California was heavily established in ranching, becoming the major supplier of cowhides and tallow to New England.[31]

The first cattle brought into the New England colonies were dairy breeds: Devon, Jersey, and Alderney—Scottish and Danish cattle. The Dutch settlement at New Amsterdam (later renamed New York by the British) was home to imported well-bred cattle. Dutch farmers in what would become New York City built log walls six feet high around their pens to protect the cattle from thieves. It became known as Wall Street. When Boston Commons was laid out in 1634, it had a twofold purpose: as a "trayning field and for the feeding of cattell."[32]

By the mid-1600s, cattle drives were common in New England. Cattle were driven to Boston from Springfield, Massachusetts, and hundreds of head were moved from the Piedmont to coastal towns. Seasonal droves of cattle were the only way to get fresh meat to market. Farther south, Kentucky cattle were driven to market at

Baltimore, while Ohio ranchers drove their animals to New England, New York, and Pennsylvania. America's first highways were "drove roads," and roadside "drove stands" catered to the drovers with food and lodging as well as feed for the cattle.

With the elements of cattle-keeping came another aspect of British highland culture, a disdain for authority. For centuries the border people in southern Scotland and northern England had fought each other. Many of the clans practiced cattle rustling, and large gangs of professional rustlers operated on both sides of the border. American cattle rustling on the Western frontier had its roots in this period and place. Continual lawlessness and violence marked the region and the people. Tenant farmers were secure in their leases because landlords relied on them to participate in fighting for protection.[33] When these people migrated to America they entered at Philadelphia and headed to the Appalachian backcountry, then into Arkansas, Missouri, Oklahoma, and Texas. In that swath of backcountry, more than half the population was from Scotland, Ireland, and northern England. Here they resumed their farming and herding, relying on family and clan support and protection. They battled the Indian residents and each other in a world of anarchy like the one they had left behind in Britain.[34]

As the country shifted its collective viewpoint, the backcountry became the frontier. People built log cabins, which Scandinavians, Germans, and north British border people had built for generations.[35] Barns and stables were crudely constructed from logs and saplings; cattle were kept in "cowpens"—simple timber fences. This "architecture of impermanence," as one historian has called it, continued in America; today's preference for mobile homes in the rural South and West echo early border settlements and their log cabins. David Hackett Fischer notes that the "mobile home is a cabin on wheels—small, cheap, simple and temporary . . . in its conception the mobile home preserves an architectural attitude that was carried to the backcountry nearly three centuries ago."[36]

From these backcountry settlers the familiar "Western" clothing style of the twentieth century developed. Women wore low-cut, tight-waisted dresses with full, short skirts; men wore shirts that were seamed horizontally across the yoke to emphasize wide shoulders. Backcountry men wore leggings and hunting shirts, just as their relatives in north Britain had done. The "Daniel Boone" look was Celtic, except for the coonskin cap, an American adaptation. Backcountry settlers located their cabins near springs or creeks, ignoring riverfront locations that might have made shipping commodities to market easier. Water, essential for keeping cattle, became particularly valuable in the arid West, and cattle-keepers fought to the death over rights to valuable springs and streams.

SIMPLY PUT, American cattle culture was divided into two camps with deep historical roots: one favored a scientific, managed, and carefully tended sort of cattle-keeping whereby profits and investments were maximized. Brought from southern Britain, established in New England, and carried into the Old Northwest and farther, it contrasted to the near-opposite cattle-keeping style of the Highlanders, who ran their unkempt animals nearly wild, made no fences, and did no selective breeding. The latter settled in the Southern pine forests and spread westward into Texas and later north into the plains states. Eventually both styles ran up against each other; neither could or would give up their traditions of cattle-keeping, even with the Civil War. Friction continued, ameliorated by the adoption of barbed wire but never completely abated. Even in today's West, where regionalism persists, it is shaped by environment to be sure, but also by tradition and heritage.

PART TWO

THE BUSINESS OF CATTLE

6

From Trails to Tracks

The ticks were there by millions—
  I tell the very truth.
For they covered up the cattle
  Just like shingles on a roof.

They sucked the very blood of life
  From the time the calves were born,
So bodies had no chance to grow
  And calf went all to horn.
    —George Jackson, Texas, 1908[1]

TEXAS LONGHORN CATTLE have been immortalized in Western literature, from 1912 when Zane Grey gave up his dental drill to write *Riders of the Purple Sage,* to Larry McMurtry's graceful Pulitzer Prize–winning novel *Lonesome Dove.* The Hollywood film industry was built largely on cowboy and Western films that featured scene after scene of rampaging herds, which brought large-scale action to the early movie screen. Today's American West is dotted with "Longhorn B-B-Q" restaurants, bars, and ephemera. These wily, wild, long-horned cattle have a cachet all their own. Their image is an icon for the American West.

Texas longhorns have also made less romantic inroads into American history. Longhorns, and their parasitic ticks, significantly altered American agriculture. Ticks infested the warm-climate herds in California and the South and led to the development of veterinary medicine within the United States Department of Agriculture, the establishment of the Bureau of Animal Industry, national quarantine laws, the Hatch Act, and eventually significant medical research into human parasitic diseases. But before the cattle tick was understood and controlled, the scourge had killed millions of cattle, financially ruined countless numbers of ranchers, and brought the livestock trade to a standstill.

Spanish cattle fever, caused by ticks, had been known in Kansas before the Civil War, and territorial laws had prevented droving infected cattle through the state. When the war put an end to longhorn drives, there was little reason to fear the fever. But immediately after the war ended, entrepreneurs headed north with Texas cattle, and Spanish fever outbreaks swept Kansas. One herd infected almost every farm along its path, leaving behind a trail of death. When infected herds passed through, local cattle, dairy cows, and even dray oxen died inexplicably. Worse, no one understood how the Texas cattle could infect the herds they encountered. No one realized that ticks fell off the Texas cattle and bit other animals. Simply tethering an animal near a Texas herd might mean death.[2]

NEW SPAIN was a great place for cattle; they multiplied quickly, and by the end of the sixteenth century there were ranches dotting the colony, from Jalisco to Durango to Chihuahua, with herds of tens of thousands. One cattle owner branded thirty thousand calves a year by century's end. But because they were so far from market, the cattle were useful only for hides, which were sent back to Spain with cargoes of silver from Mexico's mines. A few bulls were used for the ubiquitous bullfights—at a viceroy's inauguration in 1555, a Mexico City bullfight featured nearly eighty bulls. Cattle were moved up into California with the early missions and colonists; in 1769 a ship took seven head of cattle into San Diego.

Others followed, and by 1800 there were more than a million head in California.[3]

The first herds belonged to the mission ranches, large entities operated by a coalition of the Catholic church and the Spanish government. The first private ranches were land grants from the government, usually to someone with sufficient assets to put up a house and stock the land with one hundred head of cattle. The grants were usually for a league of land, about 4,440 acres. Spanish ranching culture never evolved into small operators, or family farms, unless the family was wealthy and well connected with someone in the government or monarchy.

Raising cattle was the only industry in Spanish California, and large cattle operations shaped the region's politics and social life. Similar to plantations in the American South, the large ranches were run by a patriarch and his family, passed down to the heirs, and operated by the peons, or workers, who were often no better than slaves. But California was so far from any populated markets that its beef was worthless; the cattle were valuable only for their hides and tallow, which were exported to Mexican mines or New England factories. Between 1800 and 1848, California exported five million hides. The California diet was based on beef, but unless eaten fresh, the carcass was left to rot, a practice that resulted in a huge increase in the numbers of buzzards, flies, coyotes, and bears in the areas where slaughter and skinning was done, usually about fifty cattle at a time. Spanish vaqueros, on horseback, roped and dragged the animals to the Indian laborers who did the killing, skinning, and butchering, and wrapped the tallow in the raw hides for export.

The first cattle imported into what is now Texas, about two hundred head, were brought with Spanish missionaries in 1690. The first Spanish cattle drifted off, some ran away, some were lost or stolen. There were no fenced ranges, and some stock began growing wild. In a warm climate, without many predators, herd sizes grew exponentially. Wild herds in Texas reached tens of thousands. Around 1770 the mission ranch at Goliad claimed a herd of forty thousand head. The only populated market within trading distance was Louisiana, but France, at that time a Spanish rival, controlled trade there.

Spanish and Mexican attitudes toward cattle were more businesslike than those of Celtic-British herders and farmers. The Spanish government favored large cultivated farms and preferred to eliminate pesky cattle that ate the crops. They offered bounties of four-bits (one quarter of a silver peso) a head for killing wild cattle. Thus the hunting of wild cattle became a popular sport among Spanish cavalrymen and local residents.

Wild cattle, many of which never saw a human being during their lifetime, resorted to feral behaviors. They did not move onto the Great Plains but sought cover in thickets, where they lived in small herds that grazed and watered along the rivers at night, staying under cover during the day. They always faced into the wind for scent of danger.[4] They attacked human interlopers without provocation and were unusually dangerous. Western lore is filled with stories of army columns, wagon trains, and hunting parties being attacked by wild cattle. Their speed and long sharp horns were no match for a rider on horseback, let alone afoot. They attacked horses, mules, wagons, even trees.

In 1850 a settler along the Colorado River noticed a behavioral difference between domestic and "wild" cattle. When wolves arrived in the area, his milk cows ran to the house; wild cattle in the area would form a ring around their calves, horns out, and fight the wolves off.[5] The historian Frank Dobie observes that "Cattle were more like deer than buffaloes—quick, uneasy, restless, constantly on the lookout for danger, snuffing the air and moving with a light, elastic step. In their sense of smell they were fully the equal of deer. A wounded bull has been known to hunt for his enemy by scent, trailing him on the ground like a bear."[6]

Why didn't Texas settlers capture and domesticate the wild cattle? Some tried; captured cattle were yoked alongside tame oxen to train them, but they became sullen and died. Men built sturdy pens and raised wild calves, but they did not practice long-term selective breeding. The American Southwest was a fast-moving cattle world, and few men had the time or assets to invest in quality breeding stock. Delicate breeds of cattle imported into the area were highly susceptible to disease, so the longhorn, with its immunity to Spanish fever, was a better investment.

Texas cattle ranching really began after statehood. Young Southern men who were poor but ambitious headed to the new state to work cattle. Some were paid a portion of the herd's increase each year in lieu of wages, whereby they were able to start herds of their own. By the end of the Civil War, Texas held 3.5 million cattle. Nine of ten ranch owners in Texas, however, lived in villages elsewhere. Many ranchers claimed 25,000 to 75,000 head of cattle, but they were largely absentee landlords.

For the hundreds of thousands of Texas cattle before the Civil War, there were markets only in New Orleans and Mobile. The Morgan steamship line had a monopoly on transportation between the coast of Texas and those cities, and held freight rates so high that cattle shipping was economically impossible. When the war began, Union gunboats patrolled the Mississippi River and captured New Orleans. Some herds were secretly put to swimming the Mississippi River below Vicksburg, then rushed overland to Confederate forces, but few got through.

Texas cattle were thus largely ignored—worthless without a market. Even at one or two dollars a head, there were no buyers. At that time in Texas, in fact, "a man's poverty was estimated by the number of cattle he possessed."[7] Those who had been able to sell beef to the Confederate Army had been paid with Confederate money, worthless after the war, so they found themselves "bankrupt with a cord of money."[8] By 1866, Texas cattle were overrunning the range—overgrazing the land and turning it into a desert.

In the late 1860s, however, cattle in Texas were rounded up and driven north where they brought $80 to $140 a head. In 1866, 260,000 head of cattle were herded to northern markets on speculation, but not without problems. Roving mobs of "outlaws and thieves, glad of an excuse to pillage, kill and steal," met the drovers in southern Kansas and Missouri. They robbed the drovers or stampeded the cattle. Afterward the gangs gathered the cattle for themselves or sold them back to the drover for a finder's fee. Later these predators would use the excuse of Spanish cattle fever as a pretext for such lawlessness.[9]

What was needed was a point where Northern buyers could meet Southern drovers, purchase the cattle, and ship them to Northern or Eastern markets. The railroads were extending west, but when Illinois cattle buyers tried to interest the Kansas Pacific in shipping cattle east, the railroad did not think it would be a sound venture. Generally the railroads wanted passengers; cattle were not what they had in mind. A Missouri Pacific executive ridiculed the idea, but the ambitious Hannibal and St. Joe Railroad recognized an opportunity for freight—even live cattle. Not until thousands of cattle were nearing Abilene did the Kansas Pacific agree to build a switch; even then, the tracks were laid on "cull" timbers, as executives did not think the initiative would last. They had a point—the few Texas cattle that had been shipped east were considered inedible—no one wanted to buy the meat. "Texan beef then was not considered eatable, and was as unsalable in the Eastern markets as would have been a shipment of prairie wolves," Joseph McCoy, organizer of the first drives to Abilene, wrote in 1874.[10]

Towns along the rail lines were horrified when approached by cattle shippers to build a facility for holding and shipping numbers of cattle. No town wanted anything to do with the idea, because the Texas cattle were known to carry a disease that killed other cattle. But Abilene, a gathering of a dozen "log huts" with sod roofs in 1867, had no one to protest, so investors quickly purchased land and constructed pens, barns, and offices. Abilene had grass, water, and, most valuable of all, a location as far east as drovers could go. In six months the facility held 3,000 cattle, a large Fairbanks scale, and the beginnings of a three-story hotel. In 1867, 35,000 head were driven to Abilene.

In September of that year the first rail shipment of twenty cars of cattle went from Abilene to Chicago. A railside party was held as a send-off in Abilene. Large tents were erected to receive excursionists who arrived for wine, food, and speeches. The first shipment of cattle was sold successfully in Chicago. A second found no buyers there, so the cattle were shipped farther east where they were sold for less than their freight bill. But the cattle kept coming, and by the end of 1867 a thousand rail cars of cattle had shipped out of Abilene.[11]

The melding of Southern drovers and Northern mercantilists, following a long and bloody war between the two interests, was not without problems. Texans were not eager to work with Northerners after the Civil War. Their initial experience with the southern Kansas and Missouri bandits reinforced their reluctance. But eventually the need for cooperation in getting cattle to market was a force that united Texans with their Northern business contacts. "It is true that the Western Cattle Trade has been no feeble means of bringing about an era of better feeling between Northern and Texas men by bringing them in contact with each other in commercial transactions. The feelings today [1874] existing in the breasts of all men from both sections are far different and better than they were six years ago," McCoy observed.[12]

But drovers brought trouble as well. Some sold the cattle and left without paying their bills; others were dishonest or caused trouble. "Among certain Kansans," McCoy write, "there developed an opposition as malignant as it was detestable."[13] Legislators passed the "Texas Cattle Prohibitory Law" and exerted pressure on residents of Dickinson County, where Abilene was located, to dismantle the cattle-shipping industry.

The few settlers who lived near Abilene organized to "stampede every drove of cattle that came into the county." Several Texas drovers and those who were building the shipping facility met with the locals and assured them they would profit from selling supplies to drovers, that making Abilene a shipping center would "make their county burg a head center of a great commerce, that would justly excite the envy of every rival town in the valley."[14] And they would have the advantage of buying cheap young cattle that could be fattened on hay, straw, and cornstalks—farm waste. To reinforce their promises, the party of drovers immediately bought up a big supply of butter, eggs, potatoes, onions, oats, and corn—insisting on paying double the price asked by the locals. A local farmer is reputed to have said, "Gentlemen, if I can make any money out of this cattle trade, I am not afraid of Spanish fever; but if I cannot make any money out of this cattle trade then I am d—d afraid of Spanish fever." The opposition dissolved. Not a single steer was stampeded by the locals.[15]

By the 1870s the Northern and Western "Territorial market" had evolved. Most cattle consumed on the Indian reservations of the northern plains in the 1870s came from Texas and were sold in western Kansas. Drovers competed vigorously for Indian beef contracts because the reservations bought so many cattle. The federal government purchased between thirty thousand and forty thousand head annually to feed the Indians of the upper Missouri River. Indian agency contracts were usually given to Northern brokers who bought cattle from Southern drovers. After purchasing the herds from the Texas drovers, the cattle had to be moved to the agencies in four annual waves, in order to provide fresh beef at intervals. Cattle were herded north in installments and butchered upon arrival. There they were distributed to the Indian contract beef market, where enormous profits were possible. "In those days," McCoy wrote, "an Indian contract was only another name for a big steal and swindle."[16] The beeves were sold at inflated weights; some were sold to the Indian agent one day, stolen during the night, and resold again the next day—with his complicity. Cattle brokers sought to sell to those Indian agents stationed farthest from civilization, as they were thought less likely to be "superior" officials and more willing to overlook inflated weights for a bribe.

Agency sales usually estimated or "guessed off" the weight of the beef—a fraud between the contractor and the Indian agent. It hurt the government, which paid in cash, and the Indians, who received less food than was recorded on the account books. At times there was a 50 percent disparity, according to McCoy, a gap that would have been significant to the diet of reservation Indians who were no longer hunting traditionally but relied on agency foodstuffs such as beef, flour, cornmeal, and bacon, as well as blankets, for survival.[17]

Other markets in the Territories after the Civil War were the gold mines of Idaho and Nevada, and the Union Pacific and Central Pacific railroads that were building across the continent and needed meat for their laborers. The ample supply of California cattle had disappeared years earlier, after several drought years, so the only Western supply came from Texas.[18] The demand for beef in the Western mining districts was the main reason cattle ranching spread into the far reaches of the Western territories, to bring the cattle closer to these markets. But Western mining was a boom-and-bust economy, which meant that cattle had to be moved as the market shifted location.

Colonel J. J. Myers, forty years earlier a member of John Fremont's exploring expedition in the West, was one of the few successful Texas drovers to enter the Mormon market in Utah. He sold to Salt Lake City buyers, delivering the stock overland. He trespassed his thousands of cattle over Mormon farmland, something no one could do without paying huge damage fees for the ruined crops and fences left in the herd's wake. And, it had been impossible for an outsider, a Gentile, to sell anything to the self-sufficient Mormons in 1870s Utah. Myers profited for four years, however, by informing each farm owner he met that the cattle belonged to Heber Kimball, a Mormon elder. No complaints or damage charges were ever made, according to McCoy, "and it did appear that if Heber Kimball's cattle should run over the saints bodily and tread them into the earth, it would have been all right, and not a murmur would have been heard to escape their lips." Kimball, who made a profit as the middleman, then quickly sold the cattle. "The Mormons appeared to consider it a great privilege to buy of the Sainted Elder," McCoy noted, "although they were paying from one to three dollars in gold more per head for the cattle than they would have had to pay to the Gentile drover. Indeed they would not have bought the same stock of the Gentile at any price." What Myers did not realize, however, was that his enterprise was doomed from the start. The Mormons did not purchase the cattle to butcher. By 1873 they were *exporting* cattle to the east from Utah herds, going overland to Cheyenne, then east by rail, competing in the Eastern marketplace right along with the Southerners.[19]

Driving wild cattle overland was hard work and risky. The glorified romantic tales and stories that came out of this short-lived period of American history are often centered on the cattle drive, with its stampedes, storms, and dangerous river crossings. The first few days of a drive, the men on horseback pushed the cattle hard to get them away from their home surroundings, because they tended to run off in familiar country. Drovers had to get the animals used to moving hour after hour as a herd, under the direction of men (more frequently boys) on horseback, to "break" them to the trail. The first three or four days they pushed twenty-five to thirty miles per day, then relaxed and dropped to about fifteen miles per day with ample time for the cattle to graze and water and lie down to rest two or three hours in midday. At night the herd bedded down with a few men on night watch.

A trail drive necessitated about two men for every three hundred cattle, each man using two horses so he could trade off and use a fresh mount. According to McCoy, Mexican drovers were not as favored by trail bosses, because while they worked for lower wages (if paid in gold), they were also "cruel, intolerably impudent, and mean" to the cattle.[20] This remark was probably an accurate assessment of the Mexicans' behavior; they were simply treating cattle as the Spanish had, as economic commodities. To the Texas drovers, most of Scots-Irish or West African heritage, cattle held a special esteem. They gave particular steers names, made pets of the animals now and then, and held a more caring attitude toward them. Of course these sentiments cannot be measured, but anecdotes from Texas cowboys always show their love, respect, and even admiration for the cattle.

On the trail the animals developed certain behaviors that drovers recognized and relied on. Certain cattle took the lead while others fell into their own rank in line, with certain animals trailing to the rear. Every time the herd moved out, they all went to their selected spots and began to move. A herd of a thousand stretched out from one to two miles in length and was "a very beautiful sight, inspiring the drover with enthusiasm akin to that enkindled in the breast of the military hero by the sight of marching columns of men."[21] At the rear, oxen pulled the camp wagon filled with food and medical supplies, and a wagon of calves born along the way and too young to keep up. A "cavvie yard" of extra horses made up the end of the procession. A drive into western Kansas took between twenty-five and a hundred days, depending upon the weather, the grass and water along the way, and the skill of the drover team.

Storms caused havoc, stampeding a herd in different directions from which they all had to be retrieved and brought back to the mass before moving on up the trail. To prevent stampedes, cowboys sang to the cattle in monosyllabic, tuneless songs that were simply repetitive noise to reassure the animals, like a child's lullaby. If a stampede did start, the men rode with the animals, shouting to turn the herd into a smaller and smaller circular stream. Then the men began singing loudly and the cattle calmed. The lullaby "has a great effect in quieting the herd," McCoy explained.[22]

Herds sometimes got into the habit of stampeding and became "poor and bony and get the appearance of gray hounds"—running off their weight. Running was a nearly impossible habit to break and could only be managed by putting the animals into small herds or into fenced pastures, but neither measure was very successful. It is hard to imagine today's docile, lethargic cattle running day and night, largely for the joy of it.

At night the cattle had to be settled down in a group, men or boys stationed to watch the herd in shifts. At dusk the herd lay down to "rest and ruminate"—cattle R & R. At midnight every animal rose up, turned around a few minutes, then lay back down on the alternate side for the remainder of the night. Without sentries to watch the animals, stragglers often began wandering off instead of lying down again. If that happened, it might be only a short time before the whole herd got up and moved. Wild cattle were fairly nocturnal, feeding and watering during the night. For cowboys on the trail, night was always a delicate time to maintain control of the animals.

Once the rail tracks were laid to Abilene and the infrastructure for Northern trade established, steady numbers of cattle had to be brought to market. To build up traffic on the trails out of Texas, shippers adopted a heavy advertising campaign there. Letters were sent to every "office, business house and hamlet" in Texas, stirring everyone up with "the new star of hope that had arisen in the north to light and buoy up the hitherto dark and desponding heart of the ranch man," according to McCoy. Once the Southern drovers were stimulated to bring animals north to the railheads, buyers had to be assured. Efforts were made throughout the Northern states and territories to interest buyers in the expected concentration of Southern cattle at Abilene, and in 1867, $45,000 was spent advertising in Northern newspapers. In the spring of 1868, while a crowd of buyers were ensconced in the Drover's Cottage Hotel at Abilene, waiting for the herds to arrive, Indian attacks occurred on settlements within fifty to sixty miles of the town. Word of the massacres shook the Eastern buyers: "the greatest uneasiness was manifested."[23]

With the market system in effect at Abilene, surrounding towns rushed to build corrals and entice drovers to their own sites. They hired "drummers" to meet the drovers at the Arkansas River crossing. A cash buyer for the herd was the drovers' goal, no matter which town they went to. And the buyers were there—from Illinois, Colorado, Montana, Utah, the Territories, and the Indian agencies. The cattle kept coming out of Texas; more than one thousand rail cars were filled in the month of June 1868 alone. That summer 75,000 cattle left Abilene on the railroad.

Settling arrangements in Kansas was one thing, but the cattle industry had another hurdle to overcome on the trip east. "Spanish fever," the mysterious ailment that affected nearly every Texas bovine, reared its ugly head as soon as the first herds moved eastward. When Texas cattle were sold into herds in Illinois and Indiana, within a month the domestic cattle they grazed with began to "sicken and die at a frightful rate." Scared ranchers dumped all their stock on the market before their herds sickened, which dropped beef prices.

Several exposed herds died during shipment east, causing "indignant fear and excitement" among cattlemen in both the East and West. A crash and panic hit the cattle markets and increased hostility toward Southwestern cattle that were beginning to have a terrible reputation for disease and financial ruin. In some communities, every single cow died after Texas cattle passed through. In one Illinois county only one milk cow survived. Families and subsistence farms were hardest hit, and were enraged. Anyone trying to bring Texas cattle into an area where domestic livestock grazed was likely to be "mobbed to death." The disastrous consequences of exposure to Texas cattle lay dead on the prairie; the locals near Abilene suffered a few deaths and were paid retributions by Texas drovers and cattle shippers to pacify them. In Dickinson County, where Abilene was situated, payments came to $4,500.

The Spanish fever prompted widespread alarm in the press. Owners of domestic herds across the country panicked and sold their cattle, adding falling prices to the situation. Ruin seemed inevitable. Consumers feared eating beef from the markets, not knowing whether or not it was safe. State investigations in Illinois ensued, but no cause for the cattle deaths could be determined. Cattle were quarantined, which made feedlots expensive for shippers who had to hold and feed them much longer than they had planned.

A general convention of Midwestern political officials decided that the transport of Texas cattle ought to be prohibited for eight to ten months of the year. Since the winter months were the only time the disease disappeared, that appeared to be the only safe time to move cattle up from Texas. Legislation to enforce a quarantine of the herds before moving them eastward was advocated, but regulating interstate commerce was against the Constitution, and only Congress could legislate. It became imperative to ensure cattle safety as well as supply. As McCoy observes, "The west and southwest must produce, the northwest fatten, and the east consume the beef product of the U.S.; and that one section was dependent on the other for its ultimate prosperity."[24]

In 1879 the "germ theory" of disease causation was not completely understood. One of the main theories to explain the Spanish fever was the "sporule" theory. Scientists and physicians believed that a small egg or spore was deposited on blades of grass in Texas. Texas cattle ate it, and inside their bodies the spores "disorganized" the blood and produced an illness that was contagious to other cattle. An English scientist, Professor John Gambee, fresh from tackling rinderpest in the herds of England, was brought in by the federal government to help study the matter.

It was known that tick-covered cattle somehow spread the disease, and this became the basis of another theory. Cattle coming into Illinois via the Red River route were usually heavily infested with ticks, "having so many that the actual color of the animal would be hid by the large, distended, grayish white bodies of the million of ticks which were clinging to his hide, and sucking blood from him."[25] Tick-covered cattle that entered a pasture with domestic cattle seemed to bring the sickness with them, and most hands-on cattle herders supported the tick theory. Observers had noticed that "every native [longhorn] that dies of Spanish fever will always be found to have almost one tick for every hair on his hide; that his stomach will be found often to contain ticks although small yet numerous mingled with the food."[26] Drovers had seen cattle fresh from Texas with large ticks on them; when these fell to the ground, their eggs hatched thousands of ticks. Perched on blades of grass, the ticks crawled onto the bodies of cattle as they grazed or lay down in the grass. At some point, the drovers believed, the ticks caused the animal to become sick and feverish. As proof of the tick theory, it was known that autumn's frost always ended the spread of disease on the range.

A third theory was built on the idea that Spanish fever was simply another variant of shipping fever, which affected cattle aboard steamships, the result of physical stress due to lack of rest, poor diet, and water. Since the Texas cattle had been running wild and free, they were thought to be like buffalo—likely to be upset when under restraint and handling. Cattle driven to market were put on a grueling march with near starvation rations along the way. If it was a dry season, they suffered lack of water too. It was known that if Texas cattle were herded slowly and allowed to graze a month or two before moving them in with domestic cattle, they did not cause infection.

Whatever theory one supported, there was no denying that the cattle were healthy, if rangy, in Texas; but domestic cattle that were put with them, or that grazed after them, died. Experienced cattle owners had no way to detect Spanish fever in their herds, and there was no remedy for it.

A shorthorn in pasture following Texas longhorns would, in a month or less, "become stupid, refuse to eat or drink, inclined to stand or lie in the fence corners, his head will droop below its natural position, his ears will lop down beside his head, his eyes will become nearly fixed and a wild glaring stare will be observed, whilst from his nostrils or mouth, will constantly drool a whitish ropey slobber resembling excessive salivary secretion."[27] The animal looked rough and mangy, its hair stood on end, its back arched, and it began frequent bloody urination. Searing with fever and pain, the cattle began "bellowing piercing shrieks, expressive of the racking pain. . . . Sometimes they will plunge about wildly for a few moments and then suddenly fall down and expire instantly." With milk cows it was easy to tell they were infected: their milk dried up within two days of the onset of the illness.[28]

Eventually, science and law put an end to Spanish fever, but writing in 1874, as the great cattle drives were under way, Joseph McCoy, ever a promoter, reflected optimism in spite of the dreadful disease, writing: "We have often thought that the outbreak of Spanish fever and the consequent excitement, really served to draw toward Texan cattle the attention of stock men from every quarter of the country, and eventuated in their becoming recognized as a staple commodity upon the markets."[29] Any attention at all was better than none.

But McCoy's positive spin was not enough. Eliminating the tick problem was the only way to create an export market for Texas cattle. Without access to markets, the animals were more than worthless, they were rapidly consuming pasture and breeding exponentially. Still, if you were a small-scale subsistence rancher, lack of access to Northern or Eastern markets did not matter much to you. At first, Texans were not overly concerned about Spanish fever. Southern cattle appeared to get the fever while young, and while it resulted in less weight gain and milk production, it rarely killed the animals. Because longhorns were cheaper to buy initially, a rancher's investment was so much less that the weight difference at market was not crucial. Northern cattle, bred from English stock, cost twice as much, and their meat sold for higher prices; but Texas cattle, for half the price, gained on grass more quickly than the Northern animals.[30] Thus the threat of fever to Southern cattle-raisers did not appear to be significant. Agricultural experts in the South were more interested in boosting cotton production, so little attention was given to cattle, particularly as most were being raised by subsistence farmers.

The first reaction by farmers in Missouri was to turn back Texas cattle herds for fear of passing the disease. Groups of armed men met and drove back the herds, assaulting and even killing cowboys and stampeding or killing the animals. The "Winchester quarantine" became famous as farmers met Texas cattle drives with loaded guns. An 1868 article in the *Prairie Farmer* explained that a resident "coolly loads his gun, and joins his neighbors; and they intend no scare, either. They mean to kill, do kill, and will keep killing until the drove takes the back track; and the drovers must be careful not to get between their cattle and the citizens either, unless they are bullet-proof." Some towns in Illinois prohibited the unloading of Texas cattle from trains. But some towns, such as Dodge City, thrived by allowing the animals to pass through onto railcars. The local population did not protest because few cattle were being raised in the area.[31]

Eventually states adopted quarantine laws against cattle coming from the South. Kansas prohibited cattle from below the 45th parallel from entering the state between March 1 and December 1. Other states followed suit, and by 1885 Arizona, Colorado, Montana, Nebraska, New Mexico, and Wyoming had all adopted the same legislation. Southern cattle generally were affected by tick fever, but in most of the South they were driven to local markets. The numbers and distance of their drives made Texas-Arkansas cattle so dangerous. The era of moving cheap Texas cattle north on the trails was over. Cattle could move only by train or boat, and only in winter. The result was a plunge in the price of Southern longhorn beef on the Northern market. That price had never been as high as Northern beef, though, because the longhorn meat was tough and stringy.[32]

Southern growers tried to improve their stock but found it difficult. Those who paid high prices for importing Northern herds into the South to improve blood lines lost money, for 60 to 75 percent of the animals died from Spanish fever. The fever problem also limited Southern cattle from being included in Northern stock or dairy fairs, which might have led to increasing interest in developing better breeds. The fever was a major barrier to developing modern scientific agriculture in the South.

Until the 1890s most of the investigation into Spanish fever and its eradication was pursued by the cattle raisers themselves. Then the federal government increased its role in agricultural research by establishing agricultural experiment stations. Scientific experts began to examine the fever problem, and in 1893 the USDA's Bureau of Animal Industry suggested that the cause of the disease was a parasitical protozoan which had two hosts, the cow and the tick. An infected tick could transmit the protozoan into the cow's bloodstream with a bite. This was the first theory to focus on one host animal being a vector for the infection of another. This paradigm was groundbreaking, leading eventually to the scientific understanding of human diseases such as malaria, yellow fever, and encephalitis.

Scientists also moved to find a cure. Because existing veterinary medical knowledge was limited, treatments for Spanish fever were similar to those given to humans for similar ailments. Cattle were given quinine because it was known to be effective in treating humans with malaria. Farmers were instructed to use Epsom salts, ground ginger, salt, and syrup mixtures, and to force-feed eggs and milk for high protein. Tonics that included sulfate of iron, powdered gentian root, strychnine, and salt were advised, as well as the frequent use of whiskey to reduce fever and as a "restorative." Nothing was effective.[33]

Developing an immunity to the disease carried by the tick was ultimately considered to be the best solution, and a variety of techniques were tried by scientists at the experiment stations. In one, cattle were penned up, gradually infested with ticks, and allowed a period of time to develop natural immunity. The method worked but required costly fences and pastures set aside for non-income-producing purposes, as well as hand-feeding for the animals because they could not range on pasture. Only wealthier ranchers could afford the technique. Inoculation, another method, relied on vaccinating the animals while young with vaccines made of either pulverized tick eggs, larval ticks, or blood from engorged ticks. This too was time-consuming and labor intensive. An alternative, using blood from infected cattle as a vaccine, worked, but again it was difficult to obtain the blood, and the injection was not consistent or standardized. Thinking that blood from cattle who were already immune to the disease might be the best course, researchers tried that too, but it was not effective.

In 1906, Congress funded the Department of Agriculture to support joint efforts with the fifteen fever-infested states. The initiative called for local laws to allow inspection of herds and enforce quarantine and controls over animal movement. Slowly the states passed legislation to permit federal agents to inspect herds and to provide for centers for dipping and spraying cattle. With funding reaching $250,000 in 1908, the program became one of the first federal attempts to back scientific efforts with federal funds and enforce it at the state level. Tick eradication, rather than immunization, became the focus.

Pasture rotation was the easiest way to disinfect a farm. The animals were kept in one pasture until ticks had dropped off, then were moved to another. This fit in with scientific farming's emphasis on diversified agriculture. Another practice received the seal of approval from experts: the feedlot system. This was the best way for farmers to ensure that their animals were tick free. Cattle were rotated between different feedlots every twenty days—the time it took for ticks to drop from cattle, but not enough time for eggs to hatch. After forty days, the cattle were ready to be placed in clean pasture or sent to Northern markets. The farmer could plow under or burn the pastures once the cattle had been moved out.[34]

Although these solutions to the tick problem were effective, they were not entirely acceptable because the cultural model in the Southwest was based on free-range cattle pasturing. Cattle owners would not change their long-held social customs—which helped spread Spanish fever in the first place.

Some of the tick eradication efforts caused a backlash. Scientific farming bulletins from the universities told farmers to currycomb and hand-pick the ticks from their cattle—a time-consuming idea that might engender more ridicule than adoption. How many Texas cattle owners even saw their animals periodically, let alone hand-groomed them? Another practice that created animosity was the spraying or washing of animals, adopted in 1907 by many agricultural experiment stations. Scientists stressed that the animals had to be washed or dipped in crude oil or kerosene. When performed in hot weather, this practice was sometimes devastating to the animals, and mortality among dipped animals was high. Oklahoma officials experimented with arsenic dip, which was widely adopted because it was cheaper, and the animals did not look quite so terrible after the dipping. Farmers also had problems when the stock tried to drink the dipping mixture, and some complained that dairy herds gave reduced amounts of milk afterward and that the milk was contaminated with arsenic. Too, dipping was labor-intensive and nearly impossible with large herds.

Tick eradication was tied to geography in that it had to be accomplished within an entire area. That meant it eventually had to become mandatory everywhere. When mandatory compliance went into effect, active resistance to the government, the universities, and the efforts emerged. Small-scale cattle growers resented the expense because they did not feel their animals were at risk. If they were raising Texas longhorn cattle for subsistence and local markets, the tick did not bother them. In fact, their animals were immune to it. The eradication effort aided the large-scale ranchers, who sought higher prices in the Northern rail markets.

The program also had economic consequences. Animals from below the federal quarantine line were worth $3 to $5 less than their Northern counterparts. That difference was a significant stimulus to large herd owners who could afford to construct the concrete vats necessary for dipping and the pens needed for quarantine. For them, the cost could be spread out over the herd because they were engaged in large-scale cattle operations. For growers with only a few animals, however, the cost was not feasible. This discrepancy pitted yeoman cattle owners against large-scale, capital-intensive ranchers who were the focus of government and scientific efforts. Propaganda films made by the Department of Agriculture and used to educate farmers about the tick problem portrayed the noncompliant farmers as ignorant and selfish.[35]

The solution was to require all farmers to dip cattle at regular intervals in vats built and paid for by federal, state, and county governments. Quarantine lines were established between the different jurisdictions, and cattle movement through different areas was prohibited. Moving the effort to the county level in many cases helped with compliance, but not always. Some counties ignored their responsibilities; in others, local law enforcement ignored the violence that sometimes erupted. In Lowndes and Echols counties, Georgia, in 1922, the vats and cattle pens were dynamited and burned. State and federal officials there were beaten and at least one murdered. The counties failed to intercede or prosecute the perpetrators. Farmer opposition was not quelled until the federal agents responded by mounting Browning machine guns on the vats.[36] In Texas most counties refused to cooperate with the federal program until 1917. Opposition was strongest in areas where many small-scale farmers lived. In Shelby County, farmers dynamited sixty-seven dipping vats in one night; vats throughout Brown County were dynamited in 1919.

Small-scale ranchers resented being told how to farm by wealthier competitors and scientific experts. In fact, the experts and their use of science and law eventually drove the small-scale cattle raisers out of business. Tick eradication efforts forced many small farmers to enter the cash economy by leaving the stock business and becoming sharecroppers in the expanding cotton industry. For many, what remained was a distaste for government and the meddling of scientific experts in agricultural affairs. Not until the New Deal did those suspicions change. As the historian Claire Strom notes, "Through their victory, scientists gained authority, not only in implementing their goals but also in fashioning the history of these endeavors. Resistant farmers are portrayed as stubborn and difficult to convince and are deprived of their voice as well as their livelihood."[37]

Despite Herculean efforts, not until the mid-1930s was the tick brought fairly under control. But all the effort and expense to eliminate the tick and its fever did not then increase the number of Southern cattle brought to market. For economic reasons, many due to the eradication program, fewer cattle were being raised in the South. The small family herd no longer existed. The tick program, especially the fence laws that eliminated free-range grazing, had ended that sort of cattle operation. Those growers who had owned small plots of land and relied on free use of the range could no longer utilize that resource. They had seen it coming, had known the program meant an end to their way of life. Their violent responses had reflected their lack of political power to resist what was happening to them; they could not fight government's alliance with science. As Strom explains, "Their violence was not a reflection of ignorance but represented the frustration and political expression of the marginalized."[38] They became cotton sharecroppers or left the rural South.

7

The Devil's Rope

There was no fencing on the trails to the North and we had lots of range to graze on. Now there is so much land taken up and fenced in that the trail for most of the way is little better than a crooked lane, and we have hard lines to find enough range to feed on. These fellows from Ohio, Indiana, and other northern and western states—the "bone and sinew of the country," as politicians call them—have made farms, enclosed pastures, and fenced in water holes until you can't rest; and I say D—n such bone and sinew! They are the ruin of the country, and have everlastingly, eternally, now and forever, destroyed the best grazing-land in the world.

—A nineteenth-century Texas drover[1]

WHAT OF THE BUFFALO, cattle's wild grazing cousins? After the Civil War, as railroads pushed into the heart of bison country, the animals' fate was set. Demand for thick, fluffy buffalo robes had always been steady in the Eastern and export markets, but they had to be taken from animals during the winter months when their fur was thick and full. Harsh winter conditions made hunting difficult, so most robes were taken by Indians who traded them with white merchants.

In 1870 a Philadelphia tannery developed a new technique for tanning bison hides, using chrome salts which resulted in a smooth, softer, and more pliable leather. That development sped the demise of the buffalo almost immediately. Hides, which could be taken from animals at any time of year, became more marketable than robes. Bison could now be killed in summer while the animals roamed in huge, docile herds. By the following year, commercial hide hunters had spread out over the plains, taking animals in a wild slaughter. Shooters and skinners roamed the grasslands, hauling hides to rail sidings where they were piled in stacks that looked like small hills. They were then shipped east to provide raw leather for New England's growing shoe manufacturing industry. Thick belts, made from heavy buffalo hide, were used by industry, which had moved from cog-driven to belt-driven machines. Within four years more than four million buffalo had been slaughtered on the southern plains. As the herds diminished, hunters moved westward, following them until they disappeared.

Starving and dislocated, angry Indians made their last stand during the Indian wars of the 1870s; soon they were all swept onto reservations. The vast grassy prairie went ungrazed, and a seemingly endless supply of feed for livestock sat for the taking, fueling a rush to stock the ranges with animals who would make use of it.

In 1875 the refrigerated railcar prompted a surge in cattle prices and, in the face of a devastating bout of cattle disease in England that wiped out most of the dairy and beef herds, supplied a market that was desperate for beef. Investors pushed up the price of cattle in the 1880s, and the ranges were quickly overstocked with animals. The surge of interest in Western lands brought enormous pressure to grab land before it was all gone.[2] Investors swept into the West, buying "range rights" that no one actually owned. English and Scottish cattle syndicates invested heavily, hiring administrators to watch over their interests. Interest rates ran up from 18 to 24 percent, and still there were plenty of borrowers.

With so many interests and animals crowding the range, settlers wanted to delineate and protect their property interests. Moving onto the treeless plains, they looked to adapt to the environment with familiar fencing. But Eastern fencing technologies—wooden rails, stones, levees—were impractical in open, treeless land. Boards and posts had to be shipped from Eastern or Southern forests by rail, then carried overland by ox team. Wood fences were thus expensive and difficult to build, and liable to be destroyed when prairie fires swept through. In 1871 a USDA study revealed that the cost of fencing on the plains equaled the cost of livestock, so that to fence the country would have cost more than the national debt; or more than the value of all the farm animals in the United States at the time.[3]

Hedges were one solution. Beginning in the 1860s, settlers on the plains had been planting rows of shrubbery around pastures and property lines in lieu of fences. Areas of Kansas, Iowa, and Illinois adopted hedges for nearly all fencing. The most popular hedge plant was the Osage orange, or *bois d'arc,* which had been used by Indians to make bows. Thousands of bushels of *bois d'arc* seed were exported from Texas and Arkansas nurseries into Northern states, and hedges of the treelike shrub were planted from the Illinois prairie to central Texas.[4] *Bois d'arc* hedging cost less than fifty cents a rod and grew to make a fence "pig tight, horse high, and bull strong," according to a Texas newspaper.[5]

A quart of seeds produced about five thousand plants, yet the demand was so great that the price of *bois d'arc* seeds rose to five dollars a pound. A hedge matured in four years, then the farmer had to prune the top growth so that the shrub would fill in solid below. In 1867, Kansas paid a bounty to farmers whose hedge fence was four and a half feet high and eighteen inches thick.[6] Newspapers covered all aspects of hedge-growing, feeding a frenzy which led to soaring hedge seed prices. Seeds "became the object of speculation with the usual tragic results," Walter Prescott Webb notes. Optimism soared too, and growers with visions of fortunes rushed to raise seed stock. In 1870 nearly everyone in Texas was trying to get *bois d'arc* seed either for speculation or to plant hedges.[7]

In regions where *bois d'arc* would not grow, farmers tried prickly pear, mesquite, and wild roses. But people were moving onto the plains too fast for slow-growing hedges, and hedges had a few problems: they could be destroyed by prairie fires, could not be relocated, shaded field crops, and harbored rodents that ate seeds and plants. Mostly, though, their growth was simply too slow. Americans were in a hurry.

The nineteenth century was a yeasty ferment of creativity. Fascination with inventions and patents was fueled by technological breakthroughs and the chance to make a fortune. In the first 75 years of the century, 1,200 U.S. patents for fence designs were issued, more than two-thirds of them after the Civil War. The two years between 1866 and 1868 alone saw 368 fence patents. In this wide variety of fences, made from various materials, none were successful.[8]

By the 1870s the quest for cheap, easy-to-build fencing became a near obsession on the plains. An article in a farmers' magazine observed, "It takes on the average for the whole country $1.74 worth of fences to keep $1.65 worth of stock from eating up $2.45 worth of crops." It was not long before the idea of creating an industrial replica of the *bois d'arc* branch would emerge. At the DeKalb, Illinois, county fair in 1873, Henry Rose exhibited a curious contraption he had created to keep an unruly cow from pushing through his fences. Influenced by the prickly, thorned branches of the *bois d'arc* shrub, he had taken a wooden rail and inserted sharp wire projections on it. Rose had made the prickly bar as a sort of armor: the animal actually wore it on her head, to prick her if she pushed against an existing fence. Later, after he realized it was too awkward for the cow to wear all the time, Rose decided it would work better if he attached the contraption to the top rail of the fence instead.[9]

He exhibited his fence attachment at the county fair, where it caught the attention of Joseph Glidden, a retired county sheriff and resident of DeKalb. Supposedly Glidden recognized at once the potential of the idea. He went home and created a double-stranded wire that held short barbs placed at intervals. Using a coffee grinder to twist the wire barbs, Glidden worked up a model and patented the design.

Many innovators attempted a barbed-wire design, but Glidden was the leader early on because he went beyond the design; he began fabricating wire and selling it. A farmhand who worked for Glidden described how they made wire in the evening, using the wall-mounted coffee grinder in the Glidden kitchen to make wire barbs. To twist the two lengths of wire together which would hold the barbs in place, they used a pedal-operated grindstone. They did the wire twisting in the Glidden barn, making forty-foot lengths at first. Eventually they took the process outdoors and used the DeKalb County fairgrounds. One boy climbed a windmill tower with a bucket of barbs and stuck them on the wire as the windmill brought up the wire from below. Using gravity, the barbs dropped down the wire and were spaced and twisted by workers on the ground.

Soon there were more than five hundred other wire patents and a bevy of legal battles. Michael Kelly had designed a barbed wire earlier, but Glidden was successful partly because he developed the machinery to mass-produce the product. He was challenged in court by several other inventors, and the battle continued until 1885, when the patent for barbed wire was taken to the Supreme Court. Litigation had begun in 1876, and by the time the decision was made, Glidden's initial patent had expired. In 1892 the Court declared for Glidden and sustained his rights to the patent. He had already trademarked his wire's name: "The Winner."[10]

The battle in the courts was over, but the real fight over barbed wire was on the ground. It was not immediately a hit in Texas, in fact there was a distinct lack of enthusiasm for it because it was viewed as a Northern invention. "They still suspected that this might be another Yankee scheme to benefit the industrial North at the expense of the agricultural South," the historians Henry and Frances McCallum explain. Other objections came from lumber dealers who hoped to ship fence boards west. Early-day animal rights activists got involved, too, mounting the strongest opposition to barbed wire. Even cattle raisers thought the wire cruel and protested about injuries to their animals.[11]

It took clever marketing to convince people that barbed wire was more than a passing novelty. One fellow, John Gates, epitomizes the way the new invention moved from obscurity to popularity. Gates was a wire salesman for Glidden's company. In 1876 he constructed a barbed-wire demonstration fence on a plaza in downtown San Antonio, Texas. A few longhorns were herded into the corral, and in order to showcase the wire's potential for holding cattle, the animals were frightened with torches. They spooked but could not break through the rugged, painful wire. Gates was so successful at selling the wire that he entered the manufacturing realm himself and eventually parlayed his profits into the American Steel and Wire Company and a fledgling southeast Texas oil business, the Texas Company. It eventually became Texaco.[12]

In Texas, ranchers did not use the wire to delineate enclosures but hung it in long strips to keep Northern cattle from drifting onto Southern pasture. These barriers, called "drift fences," kept free-ranging cattle out. Investors had flooded the range with cattle, and overgrazing was rampant and uncontrolled. When harsh winter weather hit the plains, with little grass left, the animals moved south—where they ran into the drift fences.

On New Year's Eve, 1885, disaster struck. A gentle snowfall turned into a howling blizzard, followed by another and another, until the plains had been hit with three violent blizzards in ten days. People were stranded or worse, and even resorted to burning their few fence posts for heat. Cattle instinctively headed south, and with harsh driving winds at their backs, they did not turn around. Pushing straight south in attempts to get ahead of the storms, they ran into the four-strand barbed-wire drift fences. Caught against the impassable wire, with more animals pushing forward from behind, the animals were trapped and perished in the driving cold. The barbed wire worked so well that none could cross. Within days thousands of frozen cattle were piled against the fences. In some places the carcasses covered an area four hundred yards from the fence. Thousands of animals died that winter, as much from the fences as the cold, in what became known as "the big die-up." Sadly, a similar series of events occurred the following winter.[13] The "Devil's Rope" had proven its impassability.

But landowners—the new farmers—continued to build fences with barbed wire. Market demand had brought competition and low prices. Nearly everyone could afford to fence their land. Barbed-wire fences favored the small landowner over the large operation. Large ranches, spread over many square miles, were the last to fence. The small operator, the farmer who sought to protect his crops, or to protect his small herd from tick infestation or unwanted breeding with the longhorns, relied on fences first. Before the widespread use of fences, the larger operations, with battalions of hired hands, waged a virtual war against small operators in many places. Powerful interests were able to run "grangers" off the land by force. But fences, once widely in place, protected small interests at the expense of the large operator who sought to move thousands of head of cattle to market. Conflicts continued, and fence-cutting led to increasing violence. The era of the Fence Cutter Wars eventually faded, but not before areas of the West were embroiled in class warfare over what form Western agriculture would take.

Oscar Flagg, a cowboy turned newspaper editor in Wyoming Territory, offers a firsthand account of what those conflicts were like during what became known as the Johnson County Cattle War. The bloody conflict pitted small landowners against an organized group of wealthy, Eastern-based entrepreneurs. Flagg went to Wyoming in 1882 as a cowboy with a herd of Texas cattle, at a time when the stock business was at its peak. "Capitalists were crowding each other to buy the golden cow," he reminisced, as herd after herd were driven into the lush, belly-high grass of the Northern plains. "Everything was prosperous, and everybody joyful."[14] At the time, most of the cattle in Wyoming were owned by large-scale English business interests who shipped them by rail to markets in Chicago at healthy profits; one London-based ranch corporation declared 35 percent annual dividends.[15]

The cattle "barons" refused to allow cowboys to purchase cattle of their own, however, and no small landowners were allowed into their domain. "There was not a small rancher between the Platte and Buffalo, or between Wind River and Little Missouri. The country was one vast feeding ground, over which their cattle grazed unrestricted. No wire fences intercepted their course from one stream to another," Flagg wrote.[16] After the harsh winter of 1885 and a summer drought, cattle perished in numbers. Beef prices fell because the animals were not in good condition. The following winter was again hard, with increasing stock losses followed by lower prices. Cowboy wage cuts resulted in a strike, which was resolved by replacing the agitators and keeping those who agreed to the lower wages. The following winter was still worse: the range had been overstocked by profiteers, and range cattle perished by the thousands from harsh weather and lack of hay. One man claimed his loss was 120 percent: "I lost all that I had, and it cost me 20 percent of what my herd was worth to find it out."[17]

The cattle barons met in Cheyenne and formed a trust in an effort to salvage what profits they could. They agreed to blackball any cowboys who had taken up land of their own, refusing to hire them for the spring roundups. Times had changed, Flagg noted; the free range of the past had been swallowed up by new arrivals. "On almost every creek large enough to furnish water for 160 acres of land could be found a rancher, with from one to fifty head of cattle. . . . Little did they dream that in the near future they would be marked for death by these barons and their homes burned over their heads."[18]

Land ownership on the plains was often difficult to determine. Large operations sometimes fenced in great areas of government-owned land that they had never laid claim to. With settlers moving to the area, and many of the employed cowboys filing 160-acre claims of their own, fences were being stapled up nearly overnight. "Some morning we will wake up to find that a corporation has run a wire fence about the boundary lines of Wyoming," the *Wyoming Sentinel* warned, "and all within the same have been notified to move on."[19] Across the West, particularly in the Northern plains where grass was plentiful and valuable land yet unclaimed, barbed wire was stitching its way in all directions. Large outfits quickly fenced in government land, daring anyone to interfere. Meddlers were warned to be "bulletproof." Later the fencing scandals would elicit government help, with President Grover Cleveland ordering federal officials to prosecute illegal fencers. But in the 1880s, Johnson County, Wyoming, had no law and order.

Large operators there ran thousands of head of cattle onto small landowners' places, tearing out fences and trampling gardens and small fields. Lynchings of the small operators followed at the hands of hired assassins. When small ranchers who had been "blackballed" by the wealthy interests sold cattle in the railyards at Omaha, their payments were seized by the Wyoming stock commission, the cattle deemed to be stolen, and the money kept by the large corporate interests. The small stockowners formed an association for their own protection and met to plan how to prosecute the big outfits.

Anger grew as the large ranches fought settlers over unmarked calves—mavericks—resorting to "blotching" the brands on branded cattle and remarking them with their own brand. Tensions reached a peak when prominent stockmen hanged James Averill and Ella Watson. Averill, a surveyor and postmaster, had detected fraud in a rancher's land claims shortly before he was hanged. Watson was a known prostitute and owner of forty head of cattle, ostensibly taken in payment for services from customers who had stolen them from the larger herds. The two were hanged without a trial, and one eyewitness, a boy, turned up missing.

The cattle interests sent recruiters to Texas, Idaho, Oregon, Arkansas, and Missouri to find men who would arm and fight the small farmers. Strangers, armed and silent, began appearing one by one in the Powder River region, and a war of intimidation began, the large outfits determined to push the settlers out of the country. Fifty mercenaries, portrayed as a surveying party, armed and outfitted with Winchester rifles, six-shooters, and horses were loaded onto railcars at Cheyenne and transported to Johnson County. If local residents were suspicious of the nighttime arrivals, they could not send a warning: telegraph lines had been cut. The illegal army set about killing settlers and torching cabins. Local residents gathered quickly and were able to hold almost a hundred of the hired assassins at a ranch house, where they assaulted them with an ingenious moving breastworks made of logs and wagon parts. Just as the settlers moved in, a military detachment interrupted the battle. The soldiers took over, the local militia was called off, and the hired thugs were taken into protective custody by the army. Shortly afterward the deputy U.S. marshal was murdered, shot in the back near Buffalo, Wyoming.

The Johnson County conflict was a tense, bitter, and divisive showdown between urban investors and rural subsistence farmers. In this case, some of the large foreign-owned operations were not even landowners, they were investors seeking profits from negligible investments. But that kind of windfall opportunity had passed. The law now called for land to be claimed, fenced, and taxed. No longer could the barons manage thousands of animals from the smoking lounge of the Cheyenne Club. The plush three-story brick building had been the center for two hundred hand-picked members who, like Sir Horace Plunkett, the EK Ranch baron, amused themselves on the club's tennis courts. With an increasing number of grangers competing for the grass, profits became elusive. When the market shifted, as it always does, the barons fought to turn back the clock to an earlier era. But economical fencing had changed everything.

Barbed wire's effects on ranching were mixed. Americans were seeking a more efficient and intelligent form of agriculture; it was no longer the domain of the free-range cattle culture. The era of the drovers, of the picturesque cattle drives, was over. While that romantic image of cattle ranching faded, the use of fences allowed ranchers to build up better herds by controlling breeding. Fences protected cattle from diseases carried by other cattle and allowed the rancher to manipulate the genetics of the herd.

Fencing generally brought a positive change for the environment too. Overgrazing could be controlled and modified by adjusting the number of animals in a pasture. Land use became more intensive, and landowners felt more responsibility for holdings they owned and maintained. Fences allowed owners of strategic natural watering holes to restrict access to neighbors' cattle. Water was an extremely valuable commodity, imperative for cattle who consumed a lot of water. Landowners with no access to natural watering holes had to invest in wells, and without electricity to power pumps, they ingeniously fashioned windmills from scrap iron, tin, and even cloth, modeled after wind-powered mills in Holland. Windmills, a wonderfully inventive means of using natural wind power, became common on the Western prairies after the land had been fenced and cross fenced. The ubiquitous "Western" windmill framed against a prairie sky has etched itself in the mind as a powerful image of the American West.

Barbed-wire fencing effectively controlled cattle, but it worked to stop people too. With the advent of "warwire" it would have another, more ominous, use. Wire was first used in combat in 1864 when smooth telegraph wire was ripped down and twisted between posts and tree stumps at Vicksburg. Barbed wire was put to military use initially in 1864 by the French army, who used both smooth and barbed wire as defensive barriers. By the 1880s barbed military wire was readily available around the world. Regular stock barbed wire was used at first, but innovations with vicious barbs made for the most difficult obstacle that troops encountered during an attack. In 1898, Teddy Roosevelt's Rough Riders used it to surround their camp during their foray in Cuba during the Spanish-American War. It was also used by the British military in the Boer War in South Africa and became a battlefield institution in Europe during World War I, when it was used to ring the front from Switzerland to the coast of Belgium, sealing the troops on either side and adding years to the trench warfare that became interminable. Thick, impenetrable lines of barbed wire froze the battles in place in a war that never seemed to end. "Trench deadlock" ensued when troops were unable to advance against the wire and faced rapid-fire machine guns. Eventually tanks that could travel over thick drifts of warwire were invented, designed to move troops through the wire barricades. The historian Paul Johnson notes that

Americans "had no idea that their invention would cost the lives of millions in World War I; had it been available in 1861, casualties in the Civil War would probably have doubled and the fighting would have prolonged itself to the end of the decade."[20]

Today barbed wire survives as a quick, economical, and effective tool for fencing cattle. It has been replaced in some pastures by battery-powered electric fences, in which a single strand can be tacked to posts. The strand gives animals an electric shock powerful enough to keep them from pushing against it. The wispy barrier is less effective in large pastures, which still rely on barbed wire. Barbed wire is widely used in urban settings now, to cap fences and walls around prisons, industrial yards, and even elementary schools. There are barbed-wire collectors and at least two barbed-wire museums, one in La Crosse, Kansas, which bills itself "The Barbed Wire Capital of the World," and another in McLean, Texas, which claims the "Largest Barbed Wire Museum in the World." The latter, named the Devil's Rope Museum, sports a monument to barbed wire made up of two huge balls of barbed wire, weighing over 370 pounds each, which sit on limestone posts. The museum is dedicated to that wire "whose existence is both absolutely beneficial to progress, at times cruel beyond comprehension, caused drastic changes in world-wide warfare, and yet protects our lives twenty-four hours each day."[21]

8

The Industrialization of Meat

Historians have deprived the packers of their rightful title of mass-production pioneers, for it was not Henry Ford but Gustavus Swift and Philip Armour who developed the assembly-line technique. . . .
—James R. Barrett, *Work and Community in the Jungle*[1]

ONE MIGHT SAY that cattle have both created and destroyed Chicago, which once billed itself as the "Great Bovine City of the World." One October evening in 1871, Mrs. O'Leary's cow reportedly kicked over a lantern, setting off a blaze that leveled the city. If the cow had a score to settle with Chicago, it would not be surprising. Chicago's Union Stockyards had become the prototype for assembly-line—rather, disassembly-line—production as the largest meatpacking facility in the world.

Getting beef to market was the challenge to which the Chicago stockyards responded. Before the coming of the railroads, driving livestock on foot to market had been the only practical way to move meat to slaughter. There was no worry the meat would spoil, for the animal could be slaughtered behind the very butcher shop in which it would be sold. Animals had to be relatively healthy in order to make the trek to market, which eliminated weak or diseased animals (they remained at home and were eaten by the farm family). And live animals were compact ways to move bulky agricultural products, such as corn or hay, in a more concentrated form—meat—to market. It was a frequent sight in any town to see small herds of animals moving through the streets.

But as towns grew in size, the practice became problematic. Driving animals along the road to market entailed noise, filth, congestion, and stress. Animal hooves were sliced and minced by the sharp cobblestones, and many animals became lame or infected on the way to market. Only those graziers who lived near a town could effectively move their livestock to market.

Railroad lines solved the problem of driving animals, and centralized both buyers and sellers. The demise of the buffalo and the rise of cattle-raising on the plains laid the foundation for Chicago's success. The environment and the economy combined to link a centrally located Chicago to the Western cattle ranges. Farmers in the Midwest had followed the farming patterns they brought from New England: multi-purpose animals raised for both meat and dairy, and a reliance on grain crops for cash income. Their farms were small, well fenced, and well tended. With the development of rail shipments of cheap Texas cattle into the market, they found it "wasteful" to feed their Midwestern cattle on grass. Land values were too high to make it pay. They switched to feeding the animals corn or grain, which worked well. Animals fattened nicely, and Midwestern meat commanded a higher price than the stringy Western beef. The next step was to eliminate the risk and expense of breeding animals and to purchase steers, "feeders," cheaply from Western producers. Those animals could be fed in a feedlot, hogs grazed along with them, and a profit made.

Chicago had been heavily involved in pork packing, like Cincinnati before it. Processing beef was the natural next step, and as demand grew, the railroads and meatpackers worked together to build a new stockyards complex. In 1864 the city's nine largest railroads joined with the meatpacking companies to build a large centralized yard, four miles from the city center—the Union Stockyards. This joint venture was a "pinnacle of Chicago's social and economic achievement, the site, above all others," according to the historian William Cronon, "that made the city an icon of nineteenth-century progress."[2] More than a million dollars went into creating a formidable operation that included 30 miles of drain pipes, 3 miles of watering troughs, and 10 miles of feed troughs. Animals were kept in 500 pens, which grew to 2,300 pens a few years later. The pens were divided into four large sections, corresponding to each of the largest rail lines. A rail line ringed the outer perimeter, making for easy transport of the animals to any sector of the yards. A six-story hotel catered to the drovers, dealers, and speculators. The Exchange Building adjoined it, where in the 1860s a bank handled half a million dollars a day. Telegraph lines overhead spread the word of prices and markets immediately. Impressive in both scale and organization, Chicago's Union Stockyards was indeed "a triumph of the engineer's craft."[3]

Here rural met urban, East met West, buyer met seller. Those who handled cattle passed them over to those who handled meat. The stockyard industrialized beef. It took animals in as a natural material, processed them into a bewildering variety of products, then sped the products off to markets. Beef became a commodity.

But meat has always had one distinct problem: it begins to rot immediately. Decay was as much the enemy of the meat industry as of the housewife and cook. Decay meant waste. It was expensive and at times dangerous to health. That was the chief reason why animals had been led alive to butcher shops and why shoppers bought fresh meat almost daily. True, salting and drying had been used for centuries, but while salted pork was delicious and widely preferred, dried and salted beef had never been as popular as fresh beef. Decay was directly related to meat temperature, which meant that most butchering had to be done in late fall or winter, when the product could be maintained at low temperature. Beef, as a market product, was local and seasonal.

As the packinghouses grew in size and scale, seasonal slaughter and processing no longer was cost-efficient. Large investments in buildings and equipment sat idle all summer when meat could not be slaughtered and shipped. A way had to be found to extend the slaughter season year-round. Climate was the problem, and eager entrepreneurs stepped in with possible solutions.

Packing meat in ice had been a natural solution but impossible before the coming of railroads. Quick rail transport made the use of ice highly effective. For decades Boston merchants had exported cakes of ice to the West Indies and Southern cities, but it had been transported by ship. Now, with rail lines probing the interior, ice was practical nearly anywhere. Ice production in winter months soared along with the building of rail lines, and by the Civil War meat producers were able to do about a third of their slaughtering in summer months by using ice to maintain the meat at cool temperatures.

The use of ice revolutionized the beef industry. Before the large-scale use of ice to pack dressed meat, the cattle that were herded up from Texas to Abilene and shipped by rail to Chicago's yards had continued their trek by rail to markets in the Eastern cities. Chicago had been merely a market, a transfer point, for most of the cattle, who went east to be slaughtered closer to the consumer. Beginning in the 1860s, following Britain's devastating cattle plague, live cattle were shipped to London by boat, packed aboard ship in a pattern picked up from slave traders a century earlier.

Chilling beef shifted the emphasis in meatpacking from pork (which had been tasty when salted) to beef (which customers preferred fresh). Cattle could be butchered in Chicago directly after sale, then shipped in cooled railcars to Eastern markets. This method would be more profitable because there would be fewer losses, common when animals weakened from the stress of long travel. Shipping rates would apply only to actual meat, not the hides, bones, and offal that made up much of the live animal's weight. No need to feed and water the animals, no more "cowpunchers" hired to poke the animals that refused to stand during transit.

In 1868 the first icebox-style railcar appeared. It was a special refrigerated car designed for fruit shipments, which packer George Hammond filled with sixteen thousand pounds of beef routed to Boston. Within five years Hammond was doing a million dollars' worth of business annually. Others quickly followed, and Gustavus Swift, a New England farm boy who had been selling meat in Boston, moved to Chicago to be closer to the cattle market. In 1875 Swift arrived in Chicago, where he purchased meat for his string of New England butcher shops. He knew about the losses connected with shipping live cattle such a distance, so he tried shipping dressed meat. He sent a few railcars east during winter months, stripping the roofs and doors off so that the meat would stay colder during transit.[4]

One problem that Swift tried to overcome was the way meat discolored if partly frozen, which happened when it came into contact with ice. Shipping the carcasses hanging from the ceiling of the railcars, so that they didn't touch ice, was difficult because the heavy beef hindquarters swayed as the car moved, sometimes tipping the railcar over on turns. His solution was to place boxes filled with ice and brine at both ends of the car, venting them so that a current of chilled air constantly flowed over the meat. This improved refrigerator car was soon copied by Swift's major competitors.[5]

Overwhelmed by the demand for ice, suppliers went farther and farther north seeking a clean, solid product. In the 1880s the Swift Company alone used nearly half a million tons of ice a year. An entire ice industry sprang up in Wisconsin, where within two decades huge structures for holding more than a hundred thousand tons of ice were in use. Workers cut, hauled, and packed the ice in sawdust. It became a major industry during the winter in an area that had relied on summer employment in farming and logging.

The logistics of shipping iced meat was a feat in itself. The meatpacking plants needed ice to cool the meat as it was butchered, and railcars needed to be packed with ice when the meat was loaded. Then, at intervals along the way, the melted ice had to be replaced with fresh frozen ice in order to maintain consistent temperatures. Swift built several icing stations along the rail line in order to resupply company shipments en route. Five stations were in place in 1883 in Michigan, Ontario, and New York state. Each railcar of meat needed about a thousand pounds of ice at each of the five stations during a four-day trip to Eastern markets. Salt, too, was important—seven hundred pounds required for each shipment.[6]

Once Swift and his cohorts got the meat to market, one more problem had to be overcome: the consumer's negative image of Western meat. People were not accustomed to purchasing beef that had been slaughtered elsewhere. How could one tell if the animal had been healthy or diseased? What might have happened to the meat during shipment? All sorts of concerns were connected with what was almost a new food product. The quickest and surest way to overcome resistance was quickly met by Chicago packers who lowered the price of their meat to about 10 percent under that of local butcher shops. Consumer resistance dissolved.

Even selling at a lower price was profitable to the Chicago packers. They bought the animals at a lower price and shipped only the meat and not the entire animal (only 55 percent of an animal was edible—the rest was wasted when shipped to the consumer). And by keeping that 45 percent of bones, hide, offal, and other inedible by-products, the packers ultimately were able to see greater profits in the marketing of what had been waste. It penciled out as a good deal for both buyer and seller.

But there was a third party involved too. The railroads had made the beef market possible in the first place, finally accepting the idea of shipping live animals over great distances. Rail lines had built stock cars and established stockyards along the rail lines, particularly in Eastern cities where they expected to continue shipping animals for slaughter. Those investments in Eastern yards would be worthless if meat was shipped east instead of live animals. So rail lines at first refused to adopt refrigerated cars, forcing the meatpackers to build and maintain their own and contract with the rail companies to move them over the tracks.

Swift, however, found a solution. One small Eastern rail line, the Grand Trunk, was willing to do whatever it took to get freight. Situated along the northern border of Lake Erie and Lake Ontario, then connecting to American railroads at Montreal, it served the Boston and New York markets. The Grand Trunk had never been able to develop much of a live cattle shipping industry because of its out-of-the-way lines and far northern route. But for refrigeration the northern line was perfect. By 1885 the Grand Trunk was hauling almost 300 million pounds of refrigerated meat annually, 60 percent of the Chicago packinghouses' output. It did not go unnoticed by the other rail lines, and it prompted a period of intense rate-cutting and secret rate negotiations between rail lines and shippers. The result was that the meatpacking companies came to control large portions of the rail industry.[7]

Eastern wholesale butchers put up a fight too, but were overcome by the Chicago packers. The wholesale butchers had fought the Chicago meat, claiming it was tainted or unsanitary, that the public's health was at stake. When the packers decided to undersell the local suppliers, they either put them out of business or allowed them to buy distribution areas for Chicago meat. The Chicago packers were ruthless competitors, and for good reason: their product was perishable, it had to be marketed quickly. Selling below cost was simply good business if the alternative was to destroy rotting meat. By selling below local prices, the packers were able to destroy competitors, and the profits from one town funded below-market prices in the next. They also interfered with Eastern butchers who tried to purchase cattle in Western markets. Cattle raisers who sold to anyone other than the Chicago firms found no market at all for their cattle the next season. In less than a decade, the Chicago packinghouses controlled the nation's beef supply.[8]

Farmers who tried to market live animals themselves, connecting with butchers on their own, discovered fewer and fewer independent butchers who would buy their animals. The only market left for them was the Chicago packinghouses, who set prices among themselves. If a farmer did not like their prices, he could go out of business or switch from raising beef cattle to dairy cattle. (Those who switched to dairying, however, would soon encounter the same corporate entities—the Chicago packinghouses—in the fight between oleomargarine and butter.)

The use of ice was revolutionary in the fight against decaying food. Chilling meat prevented bacterial action from rapidly destroying its tissues. Ice had been used by the elites in Europe since the Middle Ages but had not become widely popular as a food preservative. Most Europeans lived in rural settings or small villages, where meat could be obtained fresh. Ice had been exported from New England to the West Indies since 1805, but it was not a large industry. What made ice suddenly important was the upsurge of urban living, which removed people from sources of fresh beef and made preservation more important. Believing that disease was caused by the air, and that bad smells were connected to disease, most towns and cities forced slaughterhouses to locate far from the center of town. As cities grew, teamsters had to bring the meat from the slaughterhouse to the market.

Iced meat had another advantage in the growing marketplace: frozen or refrigerated, it could be held until it commanded higher prices. Butchers could avoid the market when there was an oversupply, holding the meat until prices rose again. It cost money to hold a live animal—for feed and water, pens and labor. With meat already butchered and hanging, or secure in barrels, it could be kept chilled and placed on the market when prices increased.

THE CATTLE PLAGUE that hit Britain and continental Europe in the 1860s wiped out the English cattle industry for years. Both beef and dairy herds were affected by the pleuropneumonia epizootic that had been around sporadically since the 1840s. In Britain all live cattle were finally ordered slaughtered by act of the government in 1868. That same year the U.S. government appointed a commission to investigate cattle diseases, because England had enacted a strict import restriction following the destruction of their herds during the epidemic. Britain's huge beef market was desperate for meat but refused to accept live animals. Imported cattle had to be slaughtered upon arrival at the docks.

The English tragedy created opportunity for those countries

where beef cattle were easily raised. With the use of iced ships, slaughtered beef could be quickly routed to the British markets from Argentina, the United States, and New Zealand. Refrigerated ships, such as the *Frigorifique*, whisked meat from Buenos Aires to France in 105 days. Ice-making machines that relied on the expansion of compressed air or the evaporation of liquid ammonia had been patented by the mid-1800s; by 1890 ships outfitted with such machinery were carrying frozen beef from Australia to London. By 1900 special ships transported millions of frozen beef carcasses from Australia, New Zealand, and Argentina. But chilled American beef, transported aboard ships packed in salt and ice, became the mainstay of the British beef market because the Atlantic voyage was shorter.[9]

As the English cattle industry recovered from the plague of the 1860s, customers there leaned heavily toward English beef. Let the poor and working classes eat the canned or chilled meat shipped in from the New World; those in the West End of London were able and willing to pay a premium for "English" beef. Undoubtedly, freshly slaughtered beef grown locally was better quality and better tasting. The animals ate local feeds, were typical English breeds rather than the rangy animals that were usually imported, and they were cut up slightly differently by butchers. Londoners *knew* English beef when they saw it and when they tasted it. At least they thought so.

Prices were higher for "English" beef in the London markets. English beef could be easily recognized in the market when compared to American beef, which had spent some time in refrigeration and was cut up differently. It did not take marketers long to see that if they could deliver live American cattle to London, they could sell them as "English" beef. But how to do it?

Cattle ships began steaming across the Atlantic in the 1880s at the same time the British home-grown beef market was reviving, giving customers a choice between fresh English beef and imported meat, either refrigerated from America or frozen from Australia and New Zealand. An investigation into the cattle-shipping industry by Samuel Plimsoll in 1890 revealed that refrigerated meat was actually the best meat on the market (the animal was killed without stress, the meat kept chilled) compared to meat from stressed, shipped cattle who had endured horrible conditions aboard ship. But cattle were being illegally shipped to England where they were slaughtered upon arrival by English butchers, cut up in the English style, and sold as "best Scotch" or "town killed" or "English" beef, at a premium price. Bringing live cattle across the Atlantic Ocean meant that middlemen, both English and American, according to Plimsoll, could profit—to the disadvantage of the English farmer and the British consumer.[10]

Cattle ships were rigged up like the Atlantic slave ships earlier in the century. Animals were penned above and below decks, given scanty amounts of food and water, and worked continually with prods, sticks, and clubs to keep them awake and alive. On delivery, shippers were paid only for live cattle, so the weakened, emaciated, and sickly animals were treated brutally to keep them standing so they could walk off the unloading dock. Hot paraffin was poured down their ear canal, hay stuffed inside their ears and set afire, and tails twisted and broken—all to incite near-dead animals to move.

The cattle ships usually carried six hundred to a thousand head of cattle and a crew of up to eighty men. The passage took about two weeks. So many animals crowded together in damp and filthy conditions created an opportune environment for infectious disease. Violent storms rocked the animals so badly that they fell and broke legs and horns, leading to infection and death. Heavily overloaded, the ships piled hay overhead, creating a top-heavy ship that was difficult to maneuver in stormy weather. Many ships went down in rough waters, taking up to a thousand animals and crew to a watery death. If a ship sank and the crew perished, widows could expect to receive their husband's pay only up to the day the ship was last seen.[11]

NOT ONLY REFRIGERATION explained the Chicago meatpackers' success: they also sold more than meat. They made use of what small butcher shops had considered waste—horns, bones, stomach lining, teeth, hooves, hides—every part of the animal was gathered up, processed, and sold by the industrial packers. Because they dealt in such volume, it was possible to ship large amounts of whatever component of the cow's body was needed by other industries: rennet, gelatin, bloodmeal, tallow, lard—the list of items made from beef by-products is almost endless. What the packers did, William Cronon points out, was to wage war on waste. Using every part of the animal conserved their resources; it was efficiency on a higher level and their most important break with the agrarian past. Philip Armour explained, "By adopting the best-known methods, nothing is wasted, and buttons, fertilizer, glue, and other things are made cheaper and better for the world in general, out of material that was before a waste and a menace."[12] No small-town butcher could market these products, which were found in small amounts in each animal. Waste was tossed in local rivers or fed to dogs or hogs. To the Chicago meatpackers, however, the by-products accounted for their profit. A steer butchered in Chicago, the meat shipped east and sold, was usually a clear loss on paper. The sale of by-products, which were essentially free to the packinghouses, made the profit. According to the packers themselves, their empires were built on waste.[13]

While the packers' use of by-products meant significantly less waste to dispose of in the Chicago area, what worried most consumers was the thought that many of these items they did not even recognize might be sold to them as foods. The public was especially nervous about the processing of adulterated or spoiled meat, which indeed went into items such as bologna sausage. Sawdust, dirt, sweepings, parts of diseased cattle—they might all be ground together and doctored with spices and potato flour. In fighting the packing interests, independent butchers claimed to be fighting for the public's health, but they were strangely silent on the issue of adulterated processed meats. After all, they had been doing it themselves for generations.

By turning meat into a commodity, the packers made it available to a greater number of working Americans. People could eat meat year round and choose from a wider range of cuts. Prices were lower than ever before. Meat had become something entirely different; people no longer associated it with a grazing steer. The package of chilled meat was no longer "alive," it was no longer warm and dripping. It was tidily cut and wrapped, and for sale along with soap and sugar. It was becoming a "man-made" product. Cowboys were reduced to romantic figures who roamed the picturesque Western ranges. That rugged, natural life was to be dreamed of, written about, and enjoyed in novels and in Buffalo Bill's Wild West Show. Roast beef, on the other hand, had become a product of the city. Nature was no longer in charge; human nature had taken over. Geography had been abstracted—it no longer mattered where the animal was born or raised. To the industrial corporation and the urban consumer, the animal itself no longer really mattered.

WHILE REFRIGERATION was a boon to providing safe meat and galvanized the beef trust's hold on the American producer and consumer, other food preservation techniques were also being developed. Canning food to preserve it from decay was invented by a Paris confectioner, François Appert, in a successful attempt to earn a cash prize offered by Napoleon in 1795. The emperor sought ways to improve the diet of the French army; Appert's method involved placing chunks of cooked meat in tin cans, adding broth, then sealing the tops with a soldered lid. The sealed cans were heated in vats of boiling water. After cooling, any cans that swelled were discarded, the rest sold. Decades later, Pasteur's groundbreaking work on the sterilization of bacteria with heat (pasteurizing) would explain how Appert's canning technique worked. Before Pasteur, no one knew why food spoiled—"spontaneous generation" (of germs) was one theory.

The English adopted the canning process in 1810 and used it to supply soup and meat to the British navy during the War of 1812. Surprisingly, canned food had little popularity in France, where it remained a military provision. In a country where fresh foods were readily available, who wanted to eat from cans?[14]

Canned foods were introduced into the United States by the British in 1817, but military and government-sponsored exploring parties were at first the only customers. In 1870 canned meat was approved as an army ration, though there had never been much need for it. Western prospectors, cowboys, and surveyors had relied on canned meats, and the U.S. navy had been buying 250 tons of the product each year. It was called canned roast beef but was actually a rather tasteless, stringy meat that had been boiled hard before processing. It was not known as a delicacy, but when they were desperate enough, people survived on it.

The huge demand generated by the Civil War and the collapse of the British cattle industry created a larger market for canned beef. But canneries were too ambitious, putting too large a portion of meat in oversize cans in an effort to gain larger sales with less labor. Heat could not penetrate evenly to the meat in the center of a large can, allowing bacteria to grow and spoiling the meat in the can or causing food poisoning when it was eaten. Not until 1900 did researchers at the Massachusetts Institute of Technology recognize the role that the acid content of the food played in eliminating bacterial growth. The level of acidity in the food determined how high the heat had to be in order to kill bacteria; higher heat and longer processing times were necessary to sterilize meat as opposed to acidic fruits like apples.

Gail Borden had made canned condensed milk popular during the Civil War and had built his business upon the nation's horror over the starving Donner Party's winter spent stranded in the Sierra Nevada Mountains. Borden liked to point out that canned food, particularly Borden's milk, would have saved the Donner Party from eventual starvation and cannibalism.

Canned food was also a socio-economic leveler in the face of the nineteenth century's growing class divisions. An official 1893 report on the benefits of canned foods read: "Preserved food has been a great democratic factor, and has nearly obliterated one of the old lines of demarcation between the poor and the wealthy. Vegetables out of season are no longer a luxury of the rich. The logger may today have a greater variety of food than could Queen Elizabeth have enjoyed with all the resources and wealth of England at her command."[15]

Chemical preservatives were also important to food preservation and in the late nineteenth century were just beginning to be included in processed foods. They were the product of research by the German dye industry that laid the foundation for a chemical revolution, eventually evolving into many different medicines. Good results in preventing spoilage in foods came from salicylic acid, sodium sulfite, benzoic acid, and formaldehyde. Chemical companies sold the preservatives to canning companies under various clever trade names but refused to label the products or tell what they were made of, as that would have allowed canners to obtain "generic" supplies. They reassured canners and consumers that the chemicals were safe and in fact a necessity. While there were few examples of people actually dying from ingesting food preservatives, there were many more accounts of people perishing from tainted food. Chemical preservatives worked. When they were added to foods, processors saw fewer cans that had to be discarded or reprocessed, or that swelled or exploded because the contents had spoiled. Because chemical preservatives worked, however, they did allow canners to process substandard foods. Although the federal government had passed a meat inspection act, that inspection took place before the animal was slaughtered. There was no inspection inside the factory or inside the can.[16]

Canned meat became a central political issue with the advent of the Spanish-American War, and nearly vaulted an unknown army officer into the White House. In February 1898 the *USS Maine* was ripped apart by a mysterious explosion while in Havana harbor, and within three months the United States had declared war on Spain. It was the first time U.S. troops had fought offshore, creating immense problems for logistics planners. Commissary officers had never provisioned for the tropics, had never had to worry about the transport and distribution of freight and supplies. The army's numbers were small, but the fighting force would have to be made up of thousands of untrained, undisciplined volunteers. Gathering them, moving them, and providing rations and supplies appeared to be a nearly insurmountable task.

Soldiers' rations were simple, scarcely changed since the Revolutionary War except for the addition of vegetables. The army had no cooks to send into the field, and no funds for hiring them, either. Legislation had to be introduced in Congress to allow the army to hire cooks, and it was opposed by conservative military leaders who thought that cooks might "pamper" American soldiers. After all, they had cooked for themselves in the field for a century—no reason to change now. The army had also failed to consider the dietary problems connected with a tropical climate. Hardtack, or "biscuits," which for decades had been shipped in multi-purpose wooden crates, grew riddled with worms in the warm, wet climate of Cuba and the Philippines. It had to be canned for the tropics. Fresh meat, long herded along behind the lines and butchered as needed, would not be possible. Canned meat would have to do.

Charles Eagan, commissary general, foresaw problems early on and attempted to solve them methodically and conservatively. Since there would be no way to provide fresh beef to the men in the field, he contracted for large amounts of canned roast beef. Most of the men had never tasted it before. Refrigerated meat could be shipped, but that would demand refrigerated ships and harbor facilities. Salted corned beef would have to make up the bulk of the ration. The War Department considered the canned roast beef to be "fresh," technically, because it was unseasoned. The navy had been using it at sea, and it was exported in large quantities for British and French troops. With a few potatoes and a carrot or onion, it made a passable stew. Eagan ordered seven million pounds of it. The order was so large that Armour recalled a substantial lot from England to fill the contract.

The men in the field—or rather aboard ship, as that was where they first sampled the canned meat—were appalled. They *hated* it. It was cold, greasy, and tasteless. Of course they were eating it cold and unsalted, straight from the can. The army had not ordered saucepans for them to cook the clever stews that Eagan had envisioned. "The beef became a torture, as the famished soldiers sickened," the historian Margaret Leech noted. "They were nauseated by the very sight of the contents of the tins. They could not choke the slimy red mess down, or could not keep it down, if they did."[17] Teddy Roosevelt, shepherding his band of volunteers called the Rough Riders, encountered one of the men in the act of throwing away his canned meat ration. "If you are a baby, you had better not come to the war. Eat it and be a man," the stalwart New Yorker turned Western rancher had admonished. When Teddy tried to eat some later, he admitted that he too was repulsed.[18]

In 1898 a small U.S. force of seventeen thousand men, including volunteers, sailed from Tampa to land in Santiago, Cuba. Two major battles were fought and won in July, and the war in Cuba quickly ended. The troops stayed on, however, "suffering nine times as many deaths from disease as from casualties in battle." The army had not been well prepared, but medical science was even more ill-prepared to deal with tropical disease and poor sanitation. Newspapers back home had a heyday with the war, from the sinking of the *Maine* to the aftermath. Reports of filthy conditions and poor food were fodder for yellow journalism that found a ready public audience. Allegations in the press so inflamed the public that President McKinley moved to appoint a special commission to investigate. The commanding general of the army, Nelson Miles, discovered in the meat an issue that both the press and the public embraced. When the war ended, Miles was quick to report to the commission that "embalmed beef," a term he was the first to use, had been foisted on his soldiers. He accused the army of having poisoned the men, of having knowingly fed the soldiers not only tainted meat but meat that had been preserved with embalming fluid. The press went wild.[19]

Indeed some bad lots *had* been sent by packinghouses eager to fill the huge army contract. "The packers," Miles charged, "foisted on the Army at least a few bad lots of the meat, cans containing scraps of gristle, pieces of rope, and even dead maggots. With complete justification, the troops loathed the product, and officers and enlisted men alike complained bitterly about its frequent issue."[20]

Miles based his charges on an eyewitness's report that the army's beef in Cuba had smelled of boric and salicylic acid—poisonous chemicals sometimes injected into meat in order to preserve it. The report came to him from a volunteer physician noted for being involved in the blackmailing of army contractors. The physician had tasted such meat before, preserved with chemicals, while on a Western hunting trip, he claimed. Miles hoped to use the beef problem as a lever into the political arena, perhaps even to put himself in the White House.[21]

The press embraced his allegations—the war helped sell newspapers, and the beef scandal revived interest. The administration's enemies also made the most of it. A painstaking investigation ensued, and eventually samples of canned beef from army supply warehouses were opened and tasted as well as examined in chemical laboratories.

Despite the commission's failure to find chemical contamination, Miles was not dissuaded and continued to press his charges. The Democrats were now considering nominating him as their presidential candidate in 1900; he even offered the vice presidential slot to Teddy Roosevelt. But Miles's sensationalism backfired. Questions were raised about why he, at the highest command of the army, had not blown the whistle earlier. If he had suspected the soldiers were being poisoned, why had he not stopped it? Why was he now attacking his own Commissary Department? The tide of public opinion now turned against him. With no proof of his allegations, it appeared that Eagan had been doing his duty, trying to procure adequate supplies at an economical expense. Dr. Walter Reed appeared before the commission and pointed out that the main cause of military mortalities had been typhoid fever, not food.[22] More than 2,000 men had died of the disease, compared to 385 who had died as a result of battle.[23]

Critics suggested that Miles had been out to get the meatpackers because he had been humiliated four years earlier during a labor strike, when his troops had been unable to control violent riots outside the Chicago packinghouses. His career folded.

But the accusations had tarnished the military as well as the canned food industry. The charges lingered, and the public continued to believe that the army had fed its men "embalmed beef." As one historian remarked, "The odor of rotten beef would always hang over the history of the war with Spain."[24]

The preservation of meat reached a new plateau with the use of radiation. As a sterilization process it was perfected during the late 1950s, but the technology simply did not catch on commercially. In February 2000 the FDA finally approved food irradiation to kill pathogens in ground beef, poultry, pork, grain, fruits, vegetables, and spices. The foods must bear a label which consists of a radura and the words "treated by irradiation." People appear to be put off by the idea of eating something they identify with radiation; only fifty percent of consumers say they would consider eating irradiated food.

While irradiation does destroy bacteria, parasites, and insects in food, it does not kill viruses (that cause hepatitis) or prions (that may cause Mad Cow disease). It does destroy some of the vitamin content and flavor.

Critics warn that using irradiation may lead to sloppy sanitation in meat processing, whereby filth is treated to make it safe to eat rather than eliminated. Supporters point out that consumers took thirty years to accept pasteurized milk and simply need time to adjust to new technology. What may make that adjustment easier is the industry's manipulation of language: the term "irradiated" food has been replaced by a new, less provocative term, "cold pasteurization." With irradiated ground beef already being sold in fifteen hundred stores, many consumers may not even know they have been eating it. Because of its quiet entrance into the marketplace, the *New York Times* calls irradiated beef patties "the first stealth food."[25]

9

Margarine: The Plastic Food

"The Truth About Oleomargarine"

Swift's Premium Oleomargarine is a sweet, pure, clean food product made from rich cream and edible fats. It contains every element of nutrition found in the best creamery butter. The process of manufacture is primitive in its simplicity, but modern in its cleanliness and purity. The butter fat in Swift's Premium Oleomargarine is microscopically and chemically the same as in the best butter; the only difference is in the way it is secured from the cow. Butter fat in butter is all obtained by churning. In Swift's Premium Oleomargarine from 1/3 to 1/2 is obtained in that way; the remainder is pressed from the choicest fat of Government-inspected animals. This pressed fat is called "Oleo," hence the name "Oleomargarine." . . . There is no coloring matter added to Premium Oleomargarine, yet it is a tempting rich cream color.

—*Kitchen Encyclopedia*, a promotional cookbook from Swift & Company, 1911[1]

OLEOMARGARINE, the most complicated and paradoxical food product in history, made its appearance in American markets at about the same time as barbed wire. Patented in the United States in 1871, it eventually prompted more legislation directed against it than any other food product in history, filling the courts with lawsuits for decades. The most totally "industrial" of any modern food product, it laid the foundation for today's highly processed convenience foods. The story behind it helps one better understand how alliances form and disintegrate in the fight over regulation—in some cases, nonregulation—of food and dairy products.

Butter has a long history. It was made in leather flasks used as horizontal churns in ancient Sumer. Later adopted by Celts and Vikings, butter-making was linked to areas of cattle culture such as France, Netherlands, Britain, and north to Iceland. Greeks and Romans ate little butter, calling it the "food of barbarians," and relied on olive oil as a dietary fat. Nondairy cultures were even suspicious of butter, some believing it caused leprosy.[2]

Butter was part of a long-running battle within the Catholic church, which banned it (along with lard and meat) on fast days and during Lent. The church sold dispensations to eat butter on fast days to those who did not wish to rely on the approved substitutes, walnut oil and olive oil. These considerable numbers of butter dispensations particularly upset Martin Luther. He criticized the church for its trafficking in butter indulgences and in 1520 printed a tract against the practice. Lines were drawn, so to speak, and interestingly, as the food historian Maguelonne Toussaint-Samat points out, those countries that traditionally used butter in their cookery were the same ones that broke away from the Catholic church in the sixteenth century. In England the king's affairs were probably more divisive, but Luther's comment that "Eating butter, they say, is a greater sin than to lie, blaspheme, or indulge in impurity" no doubt fueled popular discontent.[3] The countries of northern Europe that became Protestant are countries with a tradition of dairy farming; outsiders from Rome who insisted that they substitute olive oil (which had to be imported from

Italy, taxes paid to the church along the way) strengthened their resistance.

Worldwide, butter consumption follows linguistic patterns: English speakers eat butter while Spanish speakers prefer oil. French-speaking countries eat both. Consumers in the United States usually adopt their family's cultural heritage regarding dietary butter or oils.

In ancient religious ceremonies, butter was a sacred symbol, used as an ointment or food. The original Little Red Riding Hood tale had Red taking a pot of butter to Grandmother. Bretons believed that butter absorbed disease from the sick. In India, ceremonial butter is made from the milk of sacred cows; in Tibet it is spread on sacred statues and made into sculptures. Delicious and easy to make in the home kitchen with minimal equipment, butter has always been a favorite food. It was inevitable that with the rise of industrialism, a cheaper substitute would be sought.

Margarine, later called oleomargarine, was created in 1870 by a Frenchman from Provence, France, named Hippolyte Megé-Mouriez, in response to a prize offered by Emperor Louis Napoleon III to whomever could find a substitute for butter that could be included in army rations. Megé thought he could duplicate the way cows' bodies produced butter naturally. He thought butter was made of digested fat that was excreted through the udder, so he formulated margarine by chopping beef fat, adding pepsin (from chopped-up calves' stomach), then heating it and pressing out the oil. This he churned with chopped-up pieces of cow or hog udders and added milk, coloring, and salt. It *resembled* butter. And it was inexpensive to produce from by-products of the meat industry.[4]

Within three years Megé-Mouriez had filed for an American patent for his process and begun expanding his French margarine factory to the United States. The U.S. Dairy Company, in New York City, had already patented a copycat process, however, and was busy producing its own version of "artificial butter." Between 1870 and 1886, 180 margarine patents were applied for in the United States; like barbed wire, everyone wanted to claim a patent.

Margarine too would have effects on the nation's agriculture and economy.

By 1886 the U.S. Dairy Company had oleomargarine factories in eleven cities, producing millions of pounds of butter substitute annually. Oleomargarine had become a significant product in retail grocery markets—one with far greater profit possibilities than butter, because it was made from waste products that had been going to the soap industry. With a little tweaking, however, manufacturers discovered that margarine could be made from literally any oil fat.

Because margarine was so flexible a product, consumers worried about its safety. The nation's urbanization and new market-oriented outlook had increased competition among food producers. That competition, coupled with the public's demand for lower food prices, led manufacturers to consider the use of waste products from industry. The demand for lower prices was often met with lower-quality foods, eventually pushing manufacturers to adopt fraudulent means in order to raise profits. Foods were thinned or adulterated with lesser-quality ingredients, sometimes with waste from the processing industry itself. Milk was watered down, bread was weighted with plaster, brown sugar included sand, and meats like bologna were no more than processed offal and waste camouflaged by spices and chemicals.

No one portrays the popular attitude toward food processing—in particular the newly introduced and highly controversial product called oleomargarine—better than Mark Twain. This excerpt from *Life on the Mississippi*, published in 1883, provides insight into what the public thought about the food industry, in particular, oleomargarine:

> Speaking of manufactures reminds me of a talk upon that topic which I heard—which I overheard—on board the Cincinnati boat. . . . Two men were talking . . . it soon transpired that they were drummers . . . brisk men, energetic of movement and speech; the dollar their god, how to get it their religion.
>
> "Now as to this article," said [the one from] Cincinnati, slash-

ing into the ostensible butter and holding forward a slab of it on his knife blade, "it's from our house; look at it—smell of it—taste it. Put any test on it you want to. Take your own time—no hurry—make it thorough. There now—what do you say, butter, ain't it? Not by a thundering sight—it's oleomargarine? Yes, sir, that's what it is—oleomargarine. You can't tell it from butter; by George an expert can't! It's from our house. We supply most of the boats in the West; there's hardly a pound of butter on one of them. We are crawling right along—jumping right along is the word. We are going to have that entire trade. Yes, and the hotel trade, too. You are going to see the day, pretty soon, when you can't find an ounce of butter to bless yourself with, in any hotel in the Mississippi and Ohio valleys, outside of the big cities. Why, we are turning out oleomargarine now by the thousands of tons. And we can sell it so dirt-cheap that the whole country has got to take it—can't get around it, you see. Butter don't stand any show—there ain't any chance for competition. Butter's had its day—and from this out, butter goes to the wall. There's more money in oleomargarine than—why, you can't imagine the business we do. I've stopped in every town, from Cincinnati to Natchez; and I've sent home big orders from every one of them."

And so forth and so on, for ten minutes longer, in the same fervid strain. Then [the man from] New Orleans piped up and said:

"Yes, it's a first-rate imitation, that's a certainty; but it ain't the only one around that's first-rate. For instance, they make olive-oil out of cotton-seed oil, nowadays, so that you can't tell them apart."[5]

The two salesmen continue, one explaining to the other how his company has sold cottonseed oil as imported olive oil by putting Italian-printed labels on the bottles. Twain clearly captured the public's concern about what really was in the new commercially available foods, particularly the suspiciously cheap oleomargarines.

Twain's fictional salesmen were correct: it *was* near impossible to tell oleomargarine from butter without chemical tests, and if oleomargarine were mixed with butter, that made it still more difficult.[6] Butter and oleomargarine were both sold in bulk to the consumer. The products were scooped from large barrels, and when they sat side by side, one could not tell the difference between butter and oleomargarine unless the butter was rancid. Editors at *Scientific American* argued in 1876 that if the "made" product was so similar yet avoided the hazards of butter, such as spoilage, "it would be difficult to prove any damages save to the moral sense."

Yet the dairy industry, of course, was outraged. This substitute, this "bastard butter," directly affected the farmers' livelihood. In 1873 when the dairy industry in the Eastern states collapsed, dairy herds were slaughtered *en masse*, which weakened the already flooded beef market. Rural New England sank into a financial depression. The blame was pinned on capitalists in the Board of Trade and the Chicago meat and oleomargarine manufacturers, because "one man in a factory could make more margarine than all of the butter that all of New York's farmers could produce."[7]

State bills to levy margarine taxes sprang up to level the market, but the public continued to buy the stuff. The editor of *Puck* magazine sarcastically suggested that oleomargarine's color could be changed to identify it from butter. *Puck* called it "oily-margarine" and teased that it could be colored pink, red, or green. It was an idea quickly picked up by the anti-margarine people, and by 1880 coloring was seen as a solution.[8]

When Armour and Swift, the great Chicago packinghouses, began making a variation of margarine in the 1880s, both the patent margarine makers in the East and the nation's dairy industry were threatened. The Chicago packers had more of oleomargarine's raw material to use; Americans were eating more corn-fed beef, and there was more fat on the carcass. This beef fat was washed, minced, and pressed to remove the oil. The oil was then mixed with milk or cream to create "butterine." Lard was added to make it spread easily, but it was soon discovered that vegetable oil could substitute for lard. The use of vegetable oil brought different farmers into the game; in the South and Midwest, farmers growing oilseed plants found a market in the oleomargarine industry.

Supporters of margarine portrayed it as a battle against new technology, with backward dairy farmers trying to protect butter, a country product whose time had passed. It was a new era, after all, and margarine was part of science and progress. As Senator Zebulon Vance of North Carolina said during one congressional hearing, "If we carry our dead with us we shall travel neither far nor fast."[9] Butter, and the dairy industry, were dead. Oleomargarine was the food of the future.

Opponents of margarine used equally simple analogies—butter and margarine were *not* the same thing. The idea of margarine as a substitute for butter could be compared to feeding sawdust to a horse instead of hay; both contained cellulose but were not the same thing. Dispossessed dairy farmers angrily referred to oleomargarine as "bogus butter" or a "greasy counterfeit."[10]

Oleomargarine manufacturers enlisted science and chemistry in their cause, bringing professional scientists and their microscopes and chemistry labs into the debate. Opponents retained their own scientific professionals. Tuberculosis and trichinosis were thought to be transmitted through oleomargarine if it was not processed at sufficiently high heat. Counterarguments pointed out that butter, the "natural" product, was itself not produced in a sanitary facility. That was true: farm women had been selling butter to local merchants for decades. They brought their homemade product to the store and traded for things they wanted. The merchant dumped each woman's butter into a common barrel; when full it was shipped to the city and remixed, often with the addition of sweet cream to disguise and "freshen" the butter. Then it was sold to the consumer. The farm woman did not need to be sanitary, in fact there were no guidelines for cleanliness or the freshness of the product. Not wanting to turn away a customer, what storekeeper would reject a neighbor's butter? So everything was dumped together and sold as a fresh, natural product.

Ostensibly, margarine regulation arose to meet the public interest for two reasons. First, margarine was made from animal fat, which was a slaughterhouse waste usually reserved for soap and candle making; and it was connected with the meatpacking industry, which was viewed negatively by most everyone. Certainly animal fats rendered at low temperatures to make margarine still contained bacteria, parasitic worm eggs and larvae, and even trichinosis. Public safety was at stake. Second, margarine raised the potential for consumer fraud. If foods were sold in bulk from unlabeled barrels, how was a consumer to know if the yellow fat scooped out was butter or oleomargarine? It was impossible to tell the difference, and margarine was widely substituted for butter at both wholesale and retail levels.[11]

Moving to shut margarine out of their markets, Canadian lawmakers outlawed it in 1886; it would not be allowed in Canada again until 1949. At the same time U.S. lawmakers held numerous congressional hearings and attempted federal legislation, but multiple state regulations and court decisions preserved margarine's place in the U.S. market—except that most states adopted the tongue-in-cheek advice from *Puck* magazine: it could not be colored yellow. That restriction held until 1950. Consumers could purchase the product but had to add their own coloring. The coloring agents were eventually packaged with the margarine, and many children were given the task of kneading and coloring the margarine at home. In Canada, Ontario finally relented and allowed yellow coloring in the product in 1995; Quebec, a major dairy region, still prohibits yellow-tinted margarine.

Banning the addition of coloring to margarine was not the real issue. After all, the same natural coloring agent, annatto, was already being added to butter in winter months, when cows' diets produced butter that was off-white rather than golden. The issue was really the fact that margarine was a manufactured food, from start to finish, and thus placed dairy interests at risk.

Margarine manufacturers, on the other hand, were invariably large industrial enterprises. Margarine-making required a heavy investment in equipment to process the fat and oil, a process that did not rely on much manpower. Manufacturers fought hard and hired legal, scientific, and medical professionals to further their cause, but they suffered from their association with meatpacking.[12] By the end of the nineteenth century meatpackers had acquired a bad reputation, and the public's sympathy went to the small regional or local butter producer. It was a "David and Goliath" battle, and the public saw no need to support Armour, Swift, or Unilever, the multi-national margarine firm from Europe.

Thus phase one of the margarine war pitted the dairy industry against the big manufacturers of a profitable substitute, most of them in the meat industry. Phase two began before World War I when cottonseed oil was used to replace one-third of the meat fat in margarine. Southern cotton growers, who supplied large amounts of the oil, joined with manufacturers to resist the market controls the dairy industry sought.

Hydrogenation, the process that causes fat molecules to turn solid, was adopted during World War I. Along with the substitution of imported coconut oil, that changed the margarine playing field and its players once again. Coconut oil was used to replace 75 percent of the fat in margarine, so meatpackers and cottonseed growers now angrily challenged the imported oil as a "coconut cow" which competed with American farms. They joined the dairy industry in opposing coconut oil, calling it an "un-American" import. Several states, largely in the South and West, passed "domestic fat laws" to place taxes on margarine that did not contain domestic oils or animal fat. These states had been supporters of margarine a decade earlier when their products were incorporated in its manufacture.[13] The state fat laws caused margarine manufacturers to switch back to cottonseed oil and then, after World War II, to soybean oil. Margarine regained its popularity in the Midwest and South, where soybeans could be cultivated for the margarine industry, creating "margarine farms" that supplied the oils.[14]

As a food product, margarine continued to improve. During World War I the Danes were the first to realize the need for vitamin A when they had exported all the nation's butter, a rich source of the vitamin for most Danish children. Margarine, made from abundant fish oil in Denmark, had replaced it on the table. But an eye ailment soon developed in Danish children which they recognized as vitamin A deficiency. Vitamin A concentrates were then added to margarine to prevent the eye ailment. Realizing it could up the ante against butter by adding vitamin A during manufacturing, the U.S. margarine industry created a first for its product: an actual nutritive value.[15] This gave it a claim to nutritional status, something it had not had before. Now the customer could be assured of getting the "same" nutrition as butter, for less money.

After World War II, new technology took matters to a more complicated level. Solvent extraction was adopted to separate soybean oil from protein meal. The meal could then be processed for animal feed and the oil made into margarine. Solvent extraction was soon used in all the seed and corn oils. After the crude vegetable oil was separated by solvents, a caustic solution was added to remove free fatty acids and unwanted components. The oil was washed by mixing it with hot water, then bleached to remove its naturally dark brown color, then hydrogenated. Now a solid at room temperature, and resistant to spoiling, flavor was added along with salt to replicate the salty taste of butter.

Packaging evolved, giving customers assurance that the product was reputable—it was no longer scooped from a barrel but sold in bright, clean cartons or wrappers. The development of brand names and advertising helped move margarine from being "poor man's butter" to a product suitable for anyone interested in shopping frugally. Butter began to weaken as a status purchase. World War II and the war shortages of butter pushed margarine to the forefront of the American diet.

Following the war, dozens of popular margarine brands appeared, and the "plastic food" (according to the definition of the Food and Drug Administration) soared in popularity with a consumer population ready for manufactured foods.[16] It was a product perfect for mass marketing via the new medium of television. A food made with industrial precision, it could be designed to yield consistent results, and its appeal to buyers could be altered—it could be sold in cubes or tubs, even squirt bottles. Margarine manufacture allowed the producer to be innovative in production too, allowing for modifications in production that could increase profits.

Federal and state taxes had long been levied on margarine, to bring its price closer to butter and dissuade consumers from buying it. By 1932 margarine sales involved federal taxes and licenses for retailers who sold the product. Twenty-seven states prohibited its manufacture or sale, twenty-four states imposed a tax, and twenty-six states required state licenses for margarine sales. The army, navy, and other federal agencies were prohibited from serving margarine. Conflict over the issue of enforced pink coloring had taken the margarine battle to the Supreme Court in 1902, where a ban on the coloring of margarine to resemble butter was upheld. But the Court struck down the idea of forcing margarine manufacturers to color their product pink, which had been legislated by New Hampshire.

In 1950 the Federal Margarine Act brought an end to the federal tax on margarine. The act allowed yellow-colored margarine into the marketplace, in the dairy section, alongside its rival, butter. State bans, taxes, and restrictions fell by the wayside, and consumers gradually ate more margarine. In 1967, the last holdout, Wisconsin, "the dairy state," finally repealed its restrictions on margarine, ending the great war. But the Oleomargarine Act of 1886 had been the first federal law to regulate a domestic food product. It signaled a shift from state and local monitoring of the food supply to federal intervention and control, and laid the groundwork for further legislation.

Arguments have been made that the shift from a rural to an urban culture brought about a more liberal margarine policy in North America. But the two largest and most urbanized provinces in Canada do not fit this hypothesis. Historically, Ontario and Quebec have imposed the harshest anti-margarine laws. Both provinces—especially Quebec—are the most dairy-intensive regions in Canada.[17] In the United States, Vermont and Wisconsin have long held out protection for their dairy interests. The strongest lobby for the restrictive Oleomargarine Act of 1886 was the dairy industry of New York state, decidedly as urban as any state in the nation. Dairy producers consistently opposed margarine, but they could not ban, tax, or tint it out of existence.

IN 1998, the 125th anniversary year of margarine's first U.S. patent, a study in the British medical journal *Lancet* revealed unsettling information. Researchers in East Germany had found a link between margarine consumption and an increasing prevalence of hay fever and hypersensitivity to allergens among children. Margarine had not been consumed in East Germany before the fall of the Berlin Wall in 1989. While researchers agreed that there were probably many causes for asthma and hay fever, they found significant reasons to connect them with changes in diet, particularly the higher intake of polyunsaturated fatty acids, such as linoleic acid. While other changes in diet may have occurred, the adoption of margarine, which contains twenty times the linoleic acid of butter, was important. As the researchers, from the University Children's Hospital in Munich and Leipzig, point out, "Our data provide evidence in support of a hypothesis that deserves further investigation."[18] Opposition surfaced in a subsequent issue of *The Lancet*, as a spokesman for the Margarine Institute for Healthy Diet, in Hamburg, Germany, criticized the study.[19]

Soon other research findings exposed health risks associated with margarine consumption. The *New England Journal of Medicine* printed studies showing that risks for cardiac problems were related to margarine rather than butter, because the trans-fatty acids caused blood fats to rise. Long believed to be better than butter for arterial and cardiac health, margarine was now to blame because of its processing. Hydrogenation changes good polyunsaturates into "bad" fatty acids.

Researchers fed subjects various kinds of margarine products: soybean oil, semi-liquid margarine, soft margarine, shortening, and stick margarine. Then blood-fat levels were compared to high-butter diets. The *more* margarine eaten, the more trans-fatty acids were found in the blood, leading to *more* fats in the bloodstream. Various margarines did drive down the level of LDL cholesterol, but they also destroyed the HDL or "good" cholesterol in the bloodstream. Counting LDL versus HDL levels, butter proved to be *better* for the bloodstream than margarine. Alice Lichtenstein, a professor of nutrition at Tufts who led the research, noted, "It's stick margarine, with its high trans fatty-acid content, that is the worst offender." But other culprits were involved too. Shortening and deep-fried foods were just as bad because of hydrogenation. French fries account for three-fourths of all trans-fatty acids consumed in the United States, not because they are potatoes but because of the oils they are fried in.[20]

As margarine plummeted in popularity, no one pointed to butter as a better food. Scientists steadfastly maintained that while margarine did not lower cholesterol levels much, it did lower them a little. The next step, however, came from food manufacturers, who had cheaper and readily available alternatives waiting in the wings. Low-fat margarine, of all things, appeared in the dairy case, with gelatin as a major ingredient. In 1998 the Agricultural Research Service laboratories at the USDA were heralding a "better table spread" that would come out of the refrigerator in solid form and spread easily—without the newly recognized evil of hydrogenation. New soybean varieties had been specially bred to produce higher levels of oleic acid, a mono-unsaturated fatty acid that appears to be healthier than other oils. The government's food chemists had been working with these marvelous new soybeans, trying the old hydrogenation process as well as a new technique called interesterification, which rearranges the fat molecules to create a margarine with a higher melting point but without trans-fatty acids. Genetically modified soybeans contain 33 percent stearic acid, a saturated fat that can be processed by interesterification to make soft tub margarine.[21]

Consumers acted despite the conflicting nutritional information from advertising and government agencies. In 1999, 39 percent of consumers who had heard warnings about margarine's deficiencies stopped eating it and switched to something else; of those, 42 percent now eat butter. While the heart-health industry and medical experts decry this switch as on balance a negative development, the public seems to have decided that eating a natural product is safest after all.[22]

But margarine has not been put out to pasture just yet. Researchers at Cornell University's Department of Food Science have published a recent study that points to the safety of margarine. They note that over margarine's long history, it has never caused an incidence of food poisoning. Pathogens do not grow in margarine because it is a water-in-oil emulsion, with 2 percent salt content, that makes the growth of micro-organisms nearly impossible. Preservatives such as sorbates perform well, preventing spoilage caused by molds, which can grow in the oil portion.[23] But it is hydrogenation that keeps the margarine from spoiling and maintains its flavor. Certainly the Cornell study provides *some* justification for eating margarine, a food so sterile that no living matter can exist in it.

THROUGHOUT ITS HISTORY, margarine has sparked a running battle between shifting agricultural interests: beef, butter, cotton, and now canola or soy producers. Their organization, or lack of it, has resulted in the government's use of legislation to adjust the marketplace. The consumer has been largely absent from the debate, even though interest groups have rallied behind public-interest ideals such as food safety and consumer protection from fraud. Agricultural lobbies, with so much at stake, have kept the fight alive. But the issue essentially ended in the 1950s. Margarine, no longer "oleo" or untinted, no longer taxed or restricted, became a food in its own right, even a model for substitute margarine products—which had to be called "spreads," since they were not true margarines.

Butter, a centuries-old home-processed commodity, has not completely lost its place. In 1899 Americans ate about 20 pounds of butter per person, about 100 pounds per family. Half of it was made in the home.[24] A century later, in 1999, an average American ate 4.35 pounds of butter per year.[25] And the only butter-making the average person may encounter will be in some elementary school classrooms, where children still shake jars of whipping cream to create butter and enliven their history lessons. When the sloshing turns to thumping, and the butter appears as if by magic, amazed second-graders are just as delighted with the finished product as their forerunners were ages ago. Explaining margarine's manufacturing process to them doesn't have the same excitement.

10

Free Speech and Hamburgers

I don't know. I just—I have trouble with the cow right now. I would have trouble digesting. It's a personal thing. It has nothing to do with whether beef is safe or not safe. It's just personal.

—Oprah Winfrey, talking with Diane Sawyer after she won a court battle against Texas cattlemen, March 1998[1]

IN 1996 television audiences were shocked when newscasters announced that Oprah Winfrey, the popular Chicago television personality, was being sued for millions of dollars because she had wrinkled her pert nose and announced that she would never eat another hamburger. How could such a thing happen? What about, of all things, the First Amendment? Free speech was protected for every sort of protester, pornographer, and rabble-rouser, but not for someone who criticizes *hamburger*?

In an era when the evening newscasters unblinkingly convey details of mutilations and murder, and pry into the private lives of any and all elected officials, it seems a bit incongruous that the same newscaster might be charged with libel and brought to trial for remarking that eating a particular food is harmful. But in fact legislation to prohibit disparaging remarks about perishable agricultural products, wryly referred to as "banana bills" or "veggie libel laws," emerged as a free-speech issue in the mid-1990s.

It all began in the 1950s when researchers synthesized a growth hormone that would increase the amount of milk given by dairy cows. It was not an entirely new departure, because farmers had been inoculating dairy cows since the 1930s with hormones extracted from the pituitary glands of cattle carcasses in order to boost milk production. *Synthetic* hormones simply made the product more widely available. With the groundswell for biotechnology research, a bevy of companies jumped into bovine growth hormone research. By the mid-1980s American Cyanamid, Eli Lilly & Company, Monsanto Agricultural Company, and Upjohn all had developed recombinant bovine growth hormone (rBGH) products.

FDA officials examined research findings from the industry's trials with rBGH given to laboratory mice and ignored the weight gains the animals experienced, along with recurring conditions of diminished testicle size in male rats and diminished brain weight in female rats. Judith C. Juskevich and C. Greg Guyer, both of whom worked for the FDA's Center for Veterinary Medicine, wrote a lengthy explanation of the FDA's review of the chemical companies' lab reports (FDA did no tests of its own) which appeared in 1990 in *Science*. The FDA concluded that "the use of rBGH in dairy cattle presents no increased health risk to consumers."[2]

But the prestigious British medical journal *The Lancet* found otherwise. An article called rBST—*recombinant bovine somatotropin*, a term that the chemical companies adopted to replace "bovine growth hormone" because of negative connotations with the words *growth hormone*—"one of the most controversial drugs of the decade."[3] Canadian government health officials refused to approve the drug in that country, and about two dozen consumer groups now urge the drug be withdrawn in the United States.

The bovine drug was originally approved for use in Europe, but the European Union (EU) banned it because of its potential impact on the dairy industry and the economy. Adverse effects in cows (lameness, mastitis, and reduced life span) are cause for concern, but the debate also involves effects on humans. IGF-I, the insulin-like growth factor that results from rBST, has been linked with cell changes in the human gut, and there may be a tentative link to cancer. Because somatotropin and IGF-I appear naturally in milk, though, the FDA and the FAO/WHO (Food and Agriculture Organization/World Health Organization) both consider supplemented cows' milk to be exactly the same as natural milk.[4]

In the marketplace, American consumer response to hormone-supplemented milk has been negligible largely because producers are prohibited from labeling milk as hormone-free, an FDA ruling that protects producers who choose to use hormone supplements.

Because cows given hormone supplements produce increased amounts of milk, their bodies show the results, Howard Lyman, a fourth-generation Montana dairy and beef rancher, described his cows' experiences once he began using rBST. Milk production soared (an increase of about 20 percent), but the cows began to be stressed over the increased production: fat and body reserves drained away, and many failed to breed or had multiple births (twins or triplets). After a period of exhaustive milk production, cows actually collapsed and died from calcium loss: their bones became paper-thin. To overcome these effects on the cows, farmers began feeding specially supplemented high-protein feed. The feed was protein-rich because it was made from ground-up dead animals.

CREUTZFELDT-JAKOB DISEASE—a highly contagious disorder that affects human brain tissue—has become one of the leading health mysteries of the twenty-first century. Little is known about the disease except that it has a long incubation period (often decades), normal sterilization methods do not work, and it is always fatal. Treating patients with Creutzfeldt-Jakob disease means taking extra precautions to prevent infection. Even for simple procedures like a spinal tap, technicians must be cautioned to wear appropriate protective gear and laboratory samples must be segregated and carefully transported. Tissue samples must be submerged in an acid solution for one hour before study. All materials and instruments used with the contaminated tissues must be decontaminated and incinerated afterward to prevent the spread of infection.[5] Funeral homes refuse to embalm anyone known to have suffered from Creutzfeldt-Jakob disease; the risk of contamination from infected brain tissue is too great.[6]

Creutzfeldt-Jakob disease (CJD) was first identified in the 1920s by German scientists and occurs randomly in populations, affecting one person in a million. It has an extremely long latency period; it may not show symptoms for months, even as long as forty years. Muscle twitches, trembling, clumsiness, blurred vision, and profound dementia set in within a period of three to twelve months of onset. The symptoms are similar to those of Alzheimer's disease but accelerate much faster. Once symptoms appear, only 5 to 10 percent of patients live longer than two years.[7] Believed to be genetically based, CJD was never much of a problem because it struck so few people and so randomly. It occurs throughout the world, but little is known about how it is spread. Some people have contracted it from corneal transplants or other tissue transplants from donors unknowingly infected with CJD. Some victims have been infected after surgery using instruments previously used in brain surgery on CJD-infected patients. Some people were infected when given a human growth hormone taken from the pituitary gland of infected cadavers (synthetic growth hormone is now used). With such a long onset of symptoms, it is difficult to know when transmission has occurred.

In the 1980s cattle in Britain began succumbing to a strange sort of sickness: they paced, shook, and eventually died. Veterinarians discovered that their brain tissue was riddled with small holes, symptomatic of a condition called spongiform encephalopathy. The condition describes a group of degenerative neurological disorders which includes scrapie of sheep and goats, kuru of humans (a disease discovered among cannibals who consumed human brain tissue), chronic wasting disease of deer and elk, and transmissible encephalopathy in mink. The disease in animals has a long incubation period before the animal exhibits symptoms and eventually dies.[8] In cattle the disease is called *bovine spongiform encephalopathy* (BSE), or more commonly, Mad Cow disease. It has emerged as one of the most mysterious public health concerns in history. Formally identified in British cows in 1986, the number of cases increased rapidly. By 1997 there were 168,000 cases of BSE confirmed in Great Britain in 34,000 herds. At one point there were a thousand new cases per week.[9]

Ten years after Mad Cow disease appeared in British cattle, a new variation of Creutzfeldt-Jakob disease (called nvCJD) appeared in humans. It was different from CJD in that young people, often in their teens, began to succumb to it. The emergence of a new disease is always cause for alarm; the fact that nvCJD appeared just when the public was worried about Mad Cow disease fueled concerns by the public and the cattle industry. An important epidemiological factor is that no cases of nvCJD have been found in any geographic area free of BSE. BSE-contaminated food came to market in Britain in the period 1984–1986; the onset of the new variant cases of CJD occurred in 1994–1996. Since the identification of the new variant, dozens of cases have appeared in Great Britain and France. Although the numbers do not denote an epidemic, fears that an epidemic might be on the horizon have not disappeared. Because the disease has a long incubation period, no one knows where we stand. In twenty years, might hundreds of thousands of people succumb to nvCJD? Or will it remain an obscure infection, one or two in a million?

According to the U.S. Centers for Disease Control (CDC), no cases have been reported in the United States of BSE in cattle or nvCJD in humans. Although the CDC does no form of testing, it does study mortality data from death certificates. If a physician were to note nvCJD on the death certificate, it would alert the CDC. Worried critics point to the fact that many cases of nvCJD may be misdiagnosed as Alzheimer's disease. There is no way to determine whether a person has died from CJD or Alzheimer's without a postmortem brain biopsy.[10]

The accepted explanation for the incidence of Mad Cow disease in Britain in the 1980s is the feeding of cattle protein meal made from the rendered body parts of infected animals. Human cases of nvCJD may have occurred when people ate the infected beef.[11] It is sadly ironic that such a complicated disease is now thought to be caused by the simple practice of feeding cattle processed feedstuffs made up of the offal and body parts of dead horses and sheep, even dead cattle. But if cattle are vegetarians, why are they eating *other cattle*?

Those were nearly Oprah Winfrey's very words when she responded to revelations by Howard Lyman, the former Montana rancher and cattle feedlot owner, who appeared on her television show in April 1996. Lyman claimed that "A hundred thousand cows per year in the U.S. are fine one night, then [found] dead the following morning. The majority of those cows are ground up and fed back to other cows. If only one of them has Mad Cow disease, it has the potential to affect thousands."[12]

Shocked, Winfrey responded, "Cows are herbivores. They shouldn't be eating other cows. . . . It has just stopped me cold from eating another burger!" That comment was to cost the lively talk show host plenty of time and money, not to mention personal anxiety. As Lyman later noted wryly, "A funny thing can happen when you tell the truth in this country. You can get sued."[13]

Winfrey's program that day included two opposing views of meat industry practices. Lyman, now the program director of the National Humane Society and a cattleman turned vegetarian activist, was there to criticize the meat industry. On the same program, Gary Weber, chief scientist for the National Cattlemen's Beef Association, rebutted the claims and argued that cows were not vegetarians because they drank milk, a remark that was greeted by audience disbelief and mockery.

By June a clutch of Texas cattlemen had filed a lawsuit against Oprah and her production company, Harpo Productions, as well as

Howard Lyman, charging them with food disparagement for libeling beef. The suit claimed that Winfrey's show had defamed American beef and cost the cattle industry untold millions of dollars. The plaintiffs sought damages from Winfrey, claiming her show had been "the bomb that set it all off," according to Paul Engler, one of the plaintiffs and chief executive of Cactus Feeders of Amarillo. He claimed that the show had led to a panic in cattle trading which had caused prices to fall more than 10 percent.[14]

Could such a lawsuit be filed in a country where freedom of speech was protected by the courts and Constitution? How could anyone be sued for criticizing something as mundane as *hamburger*? For many it seemed like a publicity stunt that Winfrey might have engineered to draw attention to her show. But the charges were real, the start of a legal nightmare for Winfrey.

In the early 1990s the American Feed Industry Association, based in Alexandria, Virginia, had created a model law that they then passed on to state Farm Bureau groups, which lobbied their state legislatures for adoption.[15] Thirty-two states considered legislation that allowed agricultural producers to sue critics of their products for libel. Journalists scoffed at the measures as "veggie libel laws," but between 1991 and 1997 thirteen states adopted similar versions of the law: Alabama, Arizona, Colorado, Florida, Georgia, Idaho, Louisiana, Mississippi, North Dakota, Ohio, Oklahoma, South Dakota, and Texas. To understand where these laws originated, we can look at another American food icon: apple pie.

A 1989 television program, *60 Minutes*, had criticized the spraying of Alar, a ripening agent, on apple trees. The program linked Alar to cancer and resulted in a groundswell of public opinion against eating apples. Consumers, stores, and schools abandoned apples and apple products in droves. Even though only a portion of apple growers had been using Alar, all apple growers suffered from the downturn in the market. A group of apple growers in Washington state sued CBS, but the case was dismissed in 1993; the judge declared that the plaintiffs had not proved that the CBS report was false.

The new "veggie libel laws" were a response to that failed lawsuit. Growers claimed they needed expanded legislation because previous libel laws protected only the party named in the false statement. Other growers, although affected as a group, could not claim damages, only the named producer could. Growers and producers of agricultural products were interested in expanded legislation because their products are time-sensitive. Seasonal produce cannot be held over until publicity ebbs; negative comments may cost a seasonal industry such as fruit an entire harvest. In Texas the law was known as the False Disparagement of Perishable Food Products Act of 1995. Three lawsuits were soon filed under the legislation: one by emu growers who sued the Honda car company for making fun of emus in a television commercial; another by a sod company against a Texas agricultural agent who had criticized its product; and the suit by cattlemen against Oprah Winfrey for her comments about hamburger.[16]

At Winfrey's first trial the judge declared that cattle were not a perishable agricultural product, and she disallowed prosecution under the 1995 act. The beef industry then sued Winfrey for standard business defamation; it went on to a jury trial in Amarillo where the jurors decided in Winfrey's favor. In spite of what appeared to be a decisive victory for free speech, Winfrey's suit cost her more than $1 million in legal fees and has had larger consequences. Most important, her experience has served as a warning to the media that their remarks about farm products should be carefully considered.

The agricultural product disparagement laws seem rather innocuous, but they have had a substantial impact on the press and on consumer activists. One critic of the laws, Rodney A. Smolla, a law professor at the University of Richmond and a specialist in First Amendment issues who is concerned about censorship, explains, "It is very hard to document people who don't speak."[17] Indeed, the laws have worked effectively to eliminate criticism of agricultural products. Consumer groups have been silenced by threatening letters from organizations such as the United Fresh Fruit and Vegetable Association, which acted to cut off debate by a Vermont consumer group, Food and Water. Consumer activists now hesitate before speaking about issues related to the food supply.

Publishers too are wary of becoming test cases for the law, which no doubt will go on to higher courts before being derailed. No one wants the expense of being a test case. J. Robert Hatherill, a research scientist, had portions cut from his book *Eat to Beat Cancer* by editors who feared recriminations under the laws. Toxicologist Marc Lappé had a book canceled by an Illinois publisher after the Monsanto Company alerted the publisher to the possibility that although Illinois lawmakers had defeated perishable-product disparagement laws, the publisher could be sued under the food libel laws in other states. The publisher's libel insurance carrier dropped him, and he had no choice but to refuse to print the book. Lappé did find another publisher, and his book, *Against the Grain: Biotechnology and the Corporate Takeover of Your Food*, was published in 1998 by Common Courage Press, a small Maine publishing house. A Texas A&M University professor of journalism, Susanna Hornig Priest, found her book, *A Grain of Truth*, a book about public opinion and press coverage of the biotechnology industry, languishing with Iowa State University Press, which had contracted for the book. When Priest refused to eliminate references to the Monsanto Chemical Company, the press dropped the book. Rowman & Littlefield, a private publisher, agreed to publish it in 2001. With such difficulties in finding a public platform, even at university presses once known for academic freedom, how many other voices are not being heard?[18]

THE ALAR "INCIDENT" was explored in an October 1990 issue of *Reader's Digest* in "The Plot Against Alar" by Robert Bidinotto, the investigative journalist who had exposed Willie Horton and the Massachusetts Furlough Program to public scrutiny (and to political campaigning) in 1988. Bidinotto traced the Alar controversy to a University of Nebraska research scientist, Bela Toth, who had fed rodents massive amounts of daminozide (trade-named Alar) in the 1970s. These doses were in amounts greater than what would have been maximum tolerated doses used in cancer testing. The chemical did cause tumors in mice, but not in rats. Bidinotto points out that the Environmental Protection Agency (EPA) "disregarded the flawed study as sloppy science."[19] And Uniroyal, the maker of Alar, brought in its own scientists who "reduced Toth's studies to rubble," according to Bidinotto. The EPA decided not to ban Alar but insisted that Uniroyal obtain independent laboratory tests. Those results determined that Alar was not a carcinogen, but the EPA demanded more testing with higher levels of the chemical. At this point the rodents began to show the effects of toxic levels of the chemical. According to Bidinotto and the American Council on Science and Health, a corporate-funded research group, the dose was too high and the results were inconclusive, but the EPA proceeded to announce an Alar ban to take effect in 1990.

The EPA ban against Alar, announced in February 1989, was closely followed by the *60 Minutes* broadcast. The Alar problem did not begin with *60 Minutes*, however. Three years earlier Ralph Nader and the National Resources Defense Council (NRDC) had joined forces to work toward a ban on daminozide, using the Toth studies as their evidence. Nader's efforts at enlisting consumers in a letter-writing campaign caused Safeway Stores to drop all Alar-treated produce, which prompted other chains to follow suit. In 1988 Consumers Union, publisher of the magazine *Consumer Reports*, had conducted apple tests and found daminozide residues in raw apples and apple products.[20] New York, Connecticut, and Massachusetts had already begun to prohibit Alar-treated produce in their marketplace. Apple growers in Washington and Michigan began to halt their use of the product, and Uniroyal's market for Alar began to disappear.

Years after it has allegedly been put to rest, Alar is still part of an American dialogue. Kerry Rodgers, writing in the journal *Science, Technology & Human Values*, proposes a group-politics approach to controversy analysis. He points out that an object becomes symbolic to interest groups and turns into an invaluable political tool. Alar is such a symbol, and although the chemical itself is no longer used, it has been employed by industry, the scientific community, environmental groups, and government organizations far beyond its application as an apple-ripening agent.

The NRDC for years had been trying to change federal policies regarding agrichemicals in foods, but the release of their report on children's health risks from chemicals, *Intolerable Risk: Pesticides in Our Children's Foods*, would have gone unnoticed if not for *60 Minutes*. That show took Alar from a regulatory dispute to a national concern. Forty million viewers were told that apples, long a symbol of purity in American food, were being poisoned by American agrichemical industries. It linked apples and cancer in the public's mind. The EPA's acting administrator, John A. Moore, declared that if Alar were not already on the market, the EPA would never have allowed it for sale. No representatives from the agrichemical industry appeared on the show; Ed Bradley noted that Uniroyal had declined to appear.

The day following the broadcast, the NRDC held thirteen press conferences across the country expounding on the *Intolerable Risk* report. Apple sales plummeted as schools banned apples in lunch programs and supermarkets cleared their shelves of apple products. One consumer reportedly called an apple grower's trade association to ask if apple juice could be poured down the drain or if it should be taken to a hazardous waste collection site.[21] In November 1990 apple growers filed a lawsuit against the NRDC, *60 Minutes*, and Fenton Communications (the NRDC's public relations firm), charging that the telecast about Alar had been false and had negligently disparaged red apples. Growers asked for $250 million in damages for crops that had gone unsold due to negative publicity. They also sought to make the media more accurate in reporting about chemical use. In response, Don Hewitt of *60 Minutes* offered to air another segment explaining that Alar-treated apples posed no risk of cancer, if either the EPA or the National Cancer Institute would state such as fact. Although scientists at the institute have stated that daminozide poses less risk than the NRDC had claimed, they would not agree. A district court judge dismissed the suit; an appeals court did the same a year later.

To environmentalist-consumer groups, Alar symbolized the

government's failure to protect the nation's food supply from exposure to agrichemicals in fruits and vegetables. The entire incident created a less-than-glowing view of the EPA and presented their regulations as haphazard and inept. Ralph Nader spoke for many when he claimed in congressional testimony that statements by EPA and FDA administrators reinforced the view that government policy "pits human health and the environment against corporate profit statements."[22] The NRDC did point out that they had never intended to hurt apple growers but that the media had oversimplified their study and concentrated on apples among all produce.

Ultimately the issue shifted from a focus on FDA and Uniroyal to a fight between the NRDC and the apple growers—from a debate over health risks associated with chemical carcinogens in the diet to issues of public relations and media involvement, with *60 Minutes* bearing the brunt of the criticism. News media themselves shifted abruptly, ridiculing the entire episode and treating it as Chicken Little environmentalism and government regulation run amok. Essays, news items, and editorials repeatedly minimized the affair and even called it a false alarm. Seven years after the Alar incident, the *Columbia Journalism Review* pointed out that the threat had been real, that although viewers had overreacted to an imminent danger from Alar, the chemical substance was indeed a potent carcinogen. Even independent studies later commissioned by Uniroyal revealed a risk.

The media shift was the result of a skilled counterattack by public relations firms which attempted to portray the entire episode as a hoax. The lawsuit itself was scarcely covered in the print media. Public perception was shaped by the few offhand references to Alar that included words like "scare" and "hoax." With a lack of coverage of the issues, along with a well-orchestrated effort by industry to portray the attack on Alar as bad science, the public was unknowingly manipulated. Now, product disparagement laws have effectively silenced debate about food safety.[23] As Butch Calhoun of the Florida Fruit and Vegetable Association has said of the Florida statute, "We don't see it ever being used in Florida. We did it more as a pro-active piece of legislation."[24] In other words, the laws are intended not to take people to court for infractions but to discourage any discussion.

FREE DEBATE in a democracy is imperative, and in contemporary society it relies on the use of the media. Since the days of the Progressives and their muckraking journalism, Americans have frequently depended on investigative journalists to tell them what's going on. The Progressive press focused on such issues as municipal corruption, sweatshops and labor exploitation, child labor, monopolies, and railroad trusts. The muckrakers' "literature of exposure" covered most every conceivable kind of corruption or fraud. The middle class cheered on the journalists, and newspapers and magazine subscriptions soared along with advertising revenue. Scores of enterprising would-be muckrakers appeared on the scene, and no stone was left unturned in their eager search for scandals.[25]

At the turn of the century, patent medicine was investigated by several popular magazines and found to contain alcohol, cocaine, and opium. More shocking, the patent medicine industry was spending $40 million a year on advertising; in 1905 five patent medicine firms had each spent more than a million dollars. Advertising contracts at the time provided that the firms could pull their ads if the editorial content of the magazine or its other advertisements appeared detrimental to their interests.[26] That advertisers could influence magazine content was a surprise to many Americans at a time when so many people turned to magazines for information and acculturation.

When *Collier's Weekly* and *Ladies' Home Journal* began to expose the patent medicine industry, they were attacked by politicians, government officials, and advertisers. Both magazines refused to buckle under the pressure and took advertising revenue losses instead. Their efforts led the public and the Senate to consider legislation that would regulate medicine. The American Medical Association vehemently opposed the bill until it realized that its members would benefit financially by eliminating competition from tonics and elixirs. Even Theodore Roosevelt pushed hard for the bill, but it stalled in a House committee.

It was tales of rotten beef on America's dinner tables that finally pushed the nation into demanding action. Meatpacking was a closely controlled industry, run by a handful of large companies referred to as the "Beef Trust." Charles Edward Russell went inside the Beef Trust to write "The Greatest Trust in the World," serialized in 1905 in *Everybody's Magazine*. He not only detailed the corrupt business practices the beef industry used to keep cattle prices to producers low and meat prices to consumers high, he also noted the number of banks that had folded as cattle prices were manipulated by the monopoly. He explained how the Beef Trust controlled rail shipping, the agricultural industry, and even the electric-light and streetcar systems of Western cities.[27] Finally he explained how the beef tycoons had discovered that "preservation chemicals are cheaper than refrigeration," and that the use of boric acid, borax, formaline, and other chemicals had thus vastly increased in the processing of meat.[28] The trust's "profits are greater on inferior stock," Russell wrote. "By the use of preservatives it can make inferior meat stand transportation, and as it has no competition, and no law to fear, why should it bother?"[29]

Russell's first article in *Everybody's Magazine* prompted little reaction. A month later, with the second installment, the battle began. The Beef Trust accused him of slandering beef growers and banks in Iowa by writing about their plight. Company representatives spread out across the Western states to place advertising in local newspapers for Beef Trust products: gelatin, soap, ham, and other products—items that had "never been one-tenth so widely advertised," Russell later noted.[30] Editorials defending the beef interests appeared in newspapers across the country.

The Jungle, Upton Sinclair's novelized exposé about life in the Chicago meatpacking industry, has become a staple in American high school classes. The book, published in 1906, a year after Russell's articles, caught the American public by surprise. Other books and articles revealing the outrageous business practices of the time had already appeared, but none struck a collective nerve like *The Jungle*. The book has stayed in print for almost a century and has been translated into many foreign languages. It is hard to believe that Sinclair had trouble finding a publisher, but indeed he did.

Upton Sinclair had made a living writing pulp novels when he was offered an assignment by *The Appeal to Reason*, a socialist magazine, to go undercover and write about meatpacking methods. He worked in meat plants for seven weeks, then wrote about the filthy conditions, the bribery of inspection officials, and the use of diseased animals in food. His book-length manuscript was accepted for publication by the Macmillan Company but first appeared in serial form in the magazine. Macmillan's editors thereupon asked him to delete some of the graphic details, claiming they would damage sales. Unwilling to compromise, Sinclair took the manuscript to four other publishers—all of them turned it down. Jack London took up a collection among socialist friends to fund publication, but eventually Doubleday, Page & Company stepped in to publish the book.

As quickly as *The Jungle* appeared in print, opposition formed against it. Chicago meatpackers hired an advertising firm to write a disparaging response which was mailed to huge numbers of America's physicians, clergy, and press. Popular magazines such as the *Saturday Evening Post* attacked Sinclair and the book as sensationalist propaganda. (The *Post* was owned by a former Armour employee, who ran a series of pro-industry articles, ghosted by a staff writer, under the byline of J. Ogden Armour.) Upton Sinclair never expected such resistance from the press. He held rallies of packing plant workers and sent out reams of press releases but could not interest newspapers in covering the story. He lamented that his efforts were met with nearly complete silence.[31]

Congress held back from taking action on the much-debated Pure Food and Drug Act, which had languished in committee until the public resorted to boycotting packed meat. The pressure for reform of America's meatpacking industry climaxed when foreign countries began returning American meat and refused to import any more. The Pure Food and Drug Act, signed in the same year

The Jungle was published, required inspections as well as labeling of meat and meat products.[32]

As the public's interest in investigative reporting persisted, politicians and bureaucrats began to squirm. President Theodore Roosevelt, an adamant Progressive himself, began to criticize the press for going too far. It was TR who offered the term *muckrakers* when he said, "I want to let in light and air, but I don't want to let in sewer gas. . . . In other words, I feel that the man who in a yellow newspaper or in a yellow magazine makes a ferocious attack on good men or even attacks bad men with exaggeration for things they have not done, is a potent enemy of those of us who are really striving in good faith to expose bad men and drive them from power."[33] By 1909 the administration of William H. Taft had moved against the press, quadrupling the postal rate for periodicals, claiming it was necessary to keep the post office from losing money. Opponents argued that the rate increase was really designed to silence journalists. But the rate increase worked: subscription prices doubled, resulting in fewer subscribers and falling advertising rates.[34] Those magazines that had refused advertising from patent medicine purveyors were hardest hit by the postage increase. The Beef Trust began to squeeze the banks, whose credit was the publishing industry's lifeblood. Many publishers found themselves in the difficult position of having to end their investigative reporting or go out of business.

As magazines folded or sold out, book publishers refused to print controversial investigations. Some authors were published by socialist presses, but distributors refused to carry the books. The final straw came when the American News Company, the nation's largest magazine distributor, refused to allow muckraking magazines on their newsstands any longer. By 1910 the muckraking era was over. The muckrakers had not been able to reform the trusts simply by exposing them.[35]

Have today's media been muzzled in the debate over food safety? Certainly the agricultural product libel laws have made a difference. It is far easier to write articles concerned about the fat content of foods than about genetic alterations. But the media are simply part of the larger society, dictated by the same rules and cultural beliefs. It is only when the society itself is interested in change that the media move in that direction. In the 1830s Alexis de Tocqueville wrote that the press was imperative if democratic people wished to unite around common issues: "In democratic countries . . . it frequently happens that a great number of men who wish or who want to combine cannot accomplish it because as they are very insignificant and lost amid the crowd, they cannot see and do not know where to find one another. A newspaper then takes up the notion or the feeling that had occurred simultaneously, but singly, to each of them. All are then immediately guided towards this beacon; and these wandering minds, which had long sought each other in darkness, at length meet and unite. The newspaper brought them together, and the newspaper is still necessary to keep them united."[36]

Walter M. Brasch, a professor of journalism and specialist in media history and social issues, criticizes reporters for taking the easy way out, relying on industry-generated press releases rather than research, and on sensational crime stories for attention. He says the media wait to see what direction the public takes on issues, then follows them. We approach the next century without the Beef Trust but with a new, more sophisticated dynamic, the Biotechnology Trust. The public must engage in scientific issues, which demands efforts to understand the new technologies and what they might mean, and their historical precedents. Charles Edward Russell, writing about the Beef Trust in 1905, had good advice: "There is no remedy unless we are willing to look upon the issue as essentially an issue of morals and not of business."[37]

ULTIMATELY the strongest influence on food safety is not the media, public opinion, or even the consumer market. It is the export market. When foreign countries restrict products from their markets for food safety reasons, industry listens. In the 1990s, when the European Union and the United States refused to import British beef over fears of Mad Cow disease, the British government finally acted by destroying millions of potentially BSE-infected cattle to prevent the spread of the disease.

Such incidents have an echo in American history: the 1881 Pork War between the German government and American pork producers. When German scientists began inspecting pork to prevent trichinosis, a disease caused by parasitic roundworms found in hog muscle, Germany refused to allow American pork into the country unless it was inspected. Americans balked—food safety was not the prerogative of the government, they argued. But Chancellor Otto von Bismarck banned American pork imports. The American reaction was disbelief and anger. Newspapers criticized the Germans for what seemed like a ridiculous fear.

A study by President Chester A. Arthur revealed that American hogs did indeed harbor trichinae. Inspection bills were introduced in the 1880s, but none passed while pork exports shrank by half. Opponents charged that inspection would bring government involvement and regulation into people's lives at a nuisance level. Roger Mills, representative from Texas, claimed it would "give the President or the Secretary of Agriculture the power to control even the appetites of the people."[38] But the pork growers recognized the value of inspection: inspected meat would be welcomed in every foreign port in the world. A compromise was reached in 1890 by authorizing federal inspections of pork for export, with the agreement that if Germany refused inspected pork, German imports might also be refused. Germany dropped its import barriers the following year.

Today the United States again faces exclusion from world markets because of issues over food safety. European countries forbid the use of growth hormones in dairy cattle and refuse to allow American beef imports that have been treated with hormones. As recently as October 1999, the European Commission reaffirmed its position prohibiting imported milk or meat products from cows that receive hormone supplements.[39]

THE LANDMARK food safety acts of 1906 were pushed by public opinion from a growing national market that was increasingly urban in its mind-set. The nation was in the midst of a shift from being "working class" to becoming "consumers," a notion that spread across all classes of society. People saw food safety legislation as a triumph for the people against the special interests. They were confrontations with industrialism at a time when interest-group politics was on the rise. As James Harvey Young notes, "All major American institutions, forced into change by rapidly developing technology, had shifted from a local to a national orientation. . . . A growing agreement emerged that, since problems had become national, solutions, too must be equally encompassing. . . . Matters involving science especially seemed appropriate for such national consideration."[40] The crusade for food legislation shared with Progressive ideas a concern for "purity." Business, government, the marketplace, society, the cities, even the nation's bloodlines needed a good cleansing. The era of what would become racial hygiene was dawning.

PART THREE

THE GOOD PROVIDER

11

Breeding Back the Aurochs

It was like a miracle. The first Aurochs for
300 years could be seen alive.
—Heinz Heck, Munich Zoo,
Germany, 1921[1]

WE LIVE IN AN AGE of cloning, when the possibility of reviving long-lost forms of life, such as the dinosaurs, continues to pique our imagination. Could a laboratory somewhere actually reconstitute a tyrannosaurus from a cell fragment? And if so, what about re-creating long-lost forms of people, such as the small-brained, weak-kneed Neanderthal? Even if such feats could be accomplished, dinosaurs would never much complicate our daily lives. Neanderthals, however, might become the people next door. Your daughter might marry such a caveman throwback. Such fears are laughable now, but less than a century ago they were quite real.

THE DEVELOPMENT OF AGRICULTURE in Europe over time was uneven. As forested areas declined, pigs wandering the woods were replaced by sheep and cattle grazing on cleared pastures. As population increased, it was no longer possible to abandon worn-out land for new—and the land was wearing out quickly. Because pastures were grazed in common by everyone in the village, they were usually badly overgrazed. It was commonly remarked that five acres of private pasture were worth 250 acres of common grazing.[2] Hay was scarce, so most animals were slaughtered in the fall unless they were used for draft purposes. Winter feed was mostly straw, and by spring the animals were so weak they could barely walk.

European agriculture varied from region to region, too. Eighteenth-century French farms were still similar to those of medieval times; landholdings were small, old-fashioned hand implements were used for tending crops, and stock were not bred selectively. In contrast to easygoing French farms, English landlords adopted improvements published in agricultural and scientific papers and journals. With capital to put improvements in place, they adopted clover from Spain and turnips as a field crop from the Netherlands.[3]

Enclosing fields—the "enclosure movement" that pushed so many Scots and Irish from their homelands—was one of the new, more efficient practices, which had its roots in the Netherlands. Cattle were controlled by the fences, and their breeding could be done selectively. No longer could one renegade bull ruin the bloodlines of everyone's herd. With enclosure, bulls could be selected purposively and weaker, undesirable stock eliminated.

Today's most popular breeds of cattle originated in Britain. From Hereford, northeastern Scotland, Teesdale, and Devon, British breeds spread over the cattle-raising areas of the world: Australia, New Zealand, Europe, and North and South America. But why did these breeds originate in Britain, a small island, of all places? The answer is that two centuries ago, when breeds were being developed by selective mating, Britain was the only country where large numbers of consumers could afford to buy high-quality meat. The Industrial Revolution and world trade through colonialism had given British consumers more cash to spend. Without the potential for profit, few gains in selective breeding would have been worthwhile or even bothered with. A second advantage that Britain held was Robert Bakewell (1725–1795), a farmer in Dishley in Leicestershire. Bakewell, a tenant farmer on a piece of land just under five hundred acres, followed in his father's footsteps, trying to achieve higher-quality stock (and therefore greater returns at the market) through selective mating and ample feeding. He was born at the right time, as the swell of the Industrial Revolution created a change never before encountered in agriculture: more consumers than producers.

In the seventeenth century, half of Britain had been forest, moor, or bog—unable to provide cropland or pasture. Farming had changed little since the Roman occupation; the manorial system had divided the land into scattered military outposts that engaged in subsistence farming. As industry developed, particularly the woolen textile industry, manufacturing became linked to agriculture as the source of raw material. Farming changed to meet the new heavy demand from more urbanized industrial areas. Tenants who had worked the grounds in common, frequently overgrazing in what was known as the open-field system, were replaced as landlords consolidated their holdings into larger plots and fenced them in during the enclosure movement. Fences made possible commercial farming, which included anything that might increase the value of the land. Jethro Tull invented the mechanical seed drill, a huge improvement over hand broadcasting, which enabled more grain to be grown successfully on the same ground. The adoption of field turnips to be used as cattle feed in winter allowed more cattle to be raised on the same amount of land. Because of Tull, turnips, and Bakewell, British agriculture went from barely feeding 6.5 million people in 1750 to providing for four times that population in 1850. One of the most important factors was the improvement in livestock.[4]

Bakewell recognized the ready market for beef—consumers were tired of eating worn-out field oxen or dairy cows past their prime. He focused on creating animals that matured quickly (therefore cost less to feed) and carried tender, fatty meat (which tasted better). Bakewell worked hard to improve the land on his farm, a surprising effort since he did not own it. (When he became famous for his prize livestock, his landlord increased the rent.) Livestock feeding was an important part of his program, and he irrigated pastures and planted alternative crops to provide winter fodder. He was a firm believer in manuring his fields heavily, even taking in neighbor's animals for boarding in winter in order to obtain their manure. Manuring was controversial at the time, but Bakewell's experiments proved its value to him.[5]

Robert Bakewell is best known, however, for his practice of breeding animals selectively. It prompted both opposition and derision while bringing him fame and financial security. The Duchess of Exeter referred to him as "the Mr. Bakewell who invented sheep."[6] Animal breeding received a great deal of attention in the eighteenth century, and landlords, infused with enthusiasm for science, undertook experiments of all sorts on their farms. Agricultural research was fashionable with royalty too; George III was so enamored of his Windsor farm that he was known as "Farmer George." He conducted stock-breeding experiments and even wrote articles for agricultural journals under the name of his shepherd, "Ralph Robinson."[7] The king's interest in Bakewell and his livestock gave support and credibility to the yeoman farmer's efforts and made him popular with elite landowners who could afford to purchase or lease his stock.[8]

But Bakewell was not an elitist. He was a practical, hands-on farmer. "He was not only one with, but one *of* the rank and file of the farming community," his biographer, H. Cecil Pawson, points out.[9]

Eventually Bakewell became reticent about his methods, but examining his efforts today it is evident that he used inbreeding—breeding relatives to relatives, a practice then frowned upon as incestuous—in order to obtain animals with particular traits. His "secretive" methods depended upon both selective breeding and ample feeding. His work was "a happy marriage of art and science," which created a new breed of sheep called the Leicester—small and compactly muscled—and a smaller, intensely muscled draft horse. His beefy long-horned cattle propelled him to fame and fortune. His rams and bulls were never sold but were leased for the breeding season at handsome fees.

At a time when there were virtually no media to spread information about Bakewell's services or products, how did he gain such widespread fame and recognition? "The knowledge of Bakewell's work spread at a rate which was little short of amazing," Pawson observes. Personal recommendation, word of mouth, and the animals themselves were his advertisements. He shipped animals off to spread his name to New Zealand, Australia, Europe, and the Americas—"If they will not speak for themselves, nothing that can be said for them will do it," he said.[10] Bakewell cleverly leased his sires for a season at a time, never sold them. When animals became too old to continue breeding services, he deliberately infected them with hoof rot by putting them in flooded pastures; that way they would be purchased only for slaughter and never bred without his oversight.

Bakewell began his efforts while a young man, traveling to observe methods and livestock in the west country of England, Norfolk, Ireland, and Holland. His farm at Dishley eventually looked far different from the neighbors'. His fields were divided by thick hedges and fences (the hedges provided timber for fuel and fence rails); he dug canals (one over a mile long) to divert flood waters into ponds which were later used for pasture irrigation; he carefully grazed and rotated pastures so that they never became dry and barren. His farm was marked by tidy roads, canals for washing field turnips before feeding, and water troughs surrounded by pavement to prevent mud. In winter 170 cattle were tied indoors where an elevated floor allowed manure to be gathered and later composted and spread on the fields.

A portrait of Robert Bakewell shows him to be stocky and square-shouldered, with a wide, open countenance. He was alert to details, a careful planner, and a stickler for quality. His neighbors ridiculed him, and a contemporary, Sir Richard Phillips, wrote that "the vulgar farmers hated him." But he was honest and fair in dealings and had a bold streak. Phillips described him as a big thinker, "original," with a "playful manner, contempt of authority in thinking, and enthusiastic."[11] He identified with the practical farmer as well as the common consumer, pointing out that his philosophy was based on the fact that "you can't eat bone, therefore, give the public something to eat."[12]

The *Complete Grazier* (1833) called Bakewell "gifted with more than common acuteness of observation, judgment, and perseverance; which combined with the experience he had acquired under his father (who was also a distinguished breeder in his time) he unremittingly applied to the improvement of cattle. Such qualities directed to any one object, could not fail of success; and such it may fairly be presumed was the only mysteries he employed."[13]

François Rochefoucald, a French tourist, met Bakewell in 1774 and wrote, "He is one of the most remarkable men to meet in the whole country. . . . He set out to perfect every kind of beast that could be useful to him, and he has attained to the greatest possible measure of perfection. . . . He is now so sure of himself that he will make an offer to anyone to produce an ox for him that will put on fat on the head or the back or even in such parts of the chest or the stomach as do not usually grow fat. He even offered to make us a bet that would have some beasts that put on fat in the tail. All of this is astonishing. I do not properly understand it, but I believe it. . . . He is esteemed by the whole of England and his breeds are famous and eagerly hired."[14] Bakewell was doing the unthinkable—manipulating animals to create forms of his own choosing. "All this will seem quite astonishing and will not be believed," Rochefoucald warned, "but it is gospel truth."[15]

Dishley was more than a farm, it was an experimental research laboratory, museum classroom, and center of agricultural learning. Visitors were frequent and varied, from other farmers to elites including Russian princes and French and German dukes, and sightseers from all walks of life.[16] In the living room Bakewell kept animal skeletons as well as large glass jars of brine displaying preserved animal joints, which he used to illustrate the actual changes taking place in the anatomy of cattle bred for more muscle. He would point to the samples, telling students, "You can get beasts to weigh where you want them to weigh: in the roasting places and not the boiling places."[17]

Bakewell treated his employees as carefully as his livestock. The farm labor staff was loyal—the junior herdsman worked at Dishley for twenty years, the senior-herdsman thirty-two, and the horse superintendent forty years. Like his father, Bakewell was a member and financial supporter of the local church and actively sought visitors to the farm. His openness and willingness to show the animals and the farm along with his plainspoken demeanor and eagerness to tell stories and anecdotes lent credibility to what he was doing. His practice of leasing the animals out for breeding kept him closely connected to his clients. His business dealings were unique but highly satisfactory to everyone involved.

Before Bakewell showed that the species could be manipulated to suit man, English cattle had been categorized by their horn length: short-, middle-, or long-horned. There were more than a hundred varieties of cattle in Great Britain and Ireland, but an animal's size was the main criterion for purchase. The strength for pulling under a yoke went along with the animal's height and bulk. The Lincolnshire ox epitomized the ideal of the day: the animal stood nineteen hands high (seventy-six inches) and measured four yards from his forehead to his rump. But Bakewell's approach to animal breeding was a unique departure from what had become the norm; he sought animals that would yield the most meat rather than achieve great overall size. Contrary to the agricultural paradigm of the day, he created low-set, blocky, quick-to-mature beef cattle. He ignored the "fancy points" that the aristocratic cattle breeders had valued (mostly colorations) and looked for a way to deliver the most meat on four legs.

Bakewell's work was widely written about, causing endless speculation and comment. It created a landmark change in cattle-raising and a new era in genetic manipulation. His cattle were not the long-legged, big-boned draft oxen but a "sound, tight cylindrical carcass; wide in the hips, but very little prominence in the huckle-bones; straight back, well filled behind shoulders; neck long and fine, without any dewlap, horns long, taper downwards, and of a deep yellowish color; head fine and smooth."[18]

The effects of this new type of animal were dramatic: in 1710 the average weight of cattle at Britain's Smithfield market was 370 pounds; by 1795 the average weight was up to 800 pounds. Improved crops and feed were part of the reason, certainly, but the main influence was Robert Bakewell's innovative methods of livestock breeding.[19] Bred for meat rather than draft, cattle took on a new image. They were no longer animals that made all work possible but animals that could be manipulated in their entirety by mankind. Their value increased along with the size of their chuck roasts. The adoption of the horse collar was replacing trudging oxen with horses that worked fields at a faster pace. Steam- and coal-driven equipment that would replace horses was on the horizon. Cattle, if not selectively bred for meat by Robert Bakewell, might have passed into history at the end of the eighteenth century.

Geneticists recognize that it takes numerous animals to run a breeding program. How could Bakewell do it? He had begun with only two heifers and a bull, purchased because they were the best examples he could find. He then bred offspring and parents, destroying any animals that were not exactly as he desired. Fortunately his first calf, named "Old Comely," was remarkable—she lived twenty-six years, and at death her sirloin fat was measured at four inches thick. If Old Comely had been imperfect, Bakewell might have ended his inbreeding program as it started. But he had remarkably good luck. He overcame the limits of his farm by leasing out sires to other farmers. That allowed him to assess their progeny with different cows and in different environmental conditions, which gave him a better understanding of the sires he would use on his own herds. That way he could also see if breeding superseded environment and nutrition, allowing him to identify which traits were constant under all conditions. Keeping the best animals for his own use, and inbreeding, he was able to develop very uniform animals which passed on their desirable qualities to their progeny. Selective inbreeding has been a dominating principle in livestock improvements since Bakewell's time.[20]

In 1791 the *Introduction to a General Studbook* led the way to the recording of pedigree in horses, sheep, and cattle.[21] The short-horn was delineated a pure breed in 1822, and by 1868 the dairy breeds were becoming selective, as evidenced in the herd book of the

American Jersey Cattle Club.[22] The idea of establishing a particular identifiable "breed" was based on using selection to mate animals with similar traits so that they pass those traits on to their offspring. Those traits vary depending upon the way the animal is valued. If it is for dairy purposes, cattle with larger udders and milk production, such as Holsteins, are valued; if for beef, well-muscled hips and thighs and quick maturity are valued.

Although Bakewell died in 1795 at the age of seventy, his work did not disappear with him. One man who adopted his ideas and was shaped by them became much more famous. Charles Darwin, born nearly fifteen years after Bakewell died, based his theory of evolution due to natural selection on the foundation established at Dishley farm. Roger Wood, a zoologist at the University of Manchester, discovered many elements of Darwin's work that were directly influenced by Bakewell. "The great changes wrought by human selection within a few generations encouraged Darwin to speculate on what might be achieved by natural selection over the enormous spread of geological time," Wood notes.[23] In an essay in 1844, Darwin wrote, "Let this work of selection, on the one hand, and death on the other, go on for a thousand generations, who would pretend to affirm it would produce no effect, when we remember what in a few years Bakewell effected on cattle . . . by this identical principle of selection."[24]

The inbred lines of cattle interested Darwin most, for the new breeds that resulted showed him how quickly variations could occur. "Remember how soon Bakewell on the same principle altered cattle . . . by avoiding a cross with any breed," Darwin noted. Darwin's research was based on observation, but his reading was extensive over the years. He wrote, "I have picked up most by reading really numberless special treatises and all agricultural and horticultural journals; but it is the work of years." Just as today's researchers scrutinize sources, he added, "The difficulty is to know what to trust."[25]

Darwin's ideas jelled after reading "heaps of agricultural and horticultural books." Again and again he refers to the way "my notions about how species change are derived from long continued study of the works of . . . agriculturalists and horticulturalists."[26] He calls Bakewell "the famous Bakewell" and admires the "skill and perseverance shown by the men who have left an enduring monument of their success on the present state of domesticated animals. . . ." As Wood notes, "We can be certain that he had Robert Bakewell very much in the forefront of his mind."[27]

IT DID NOT TAKE LONG for the idea of creating a particular style of cow to evolve into the idea that perhaps people too could be "improved." The study of genetics had begun in fits and starts. Gregor Mendel, an Austrian monk, had read Darwin's *Origin of Species* and other horticultural works and experimented with garden peas and their genetic heritage. He published his findings, *Experiments with Plant Hybrids*, in 1867, in a scientific paper that languished for forty years after its publication. Not until the beginning of the twentieth century, as the mood for creating clean, healthy environments took hold, was his paper pulled from the dustbin of literature and recognized as a landmark scientific discovery, laying the foundation for modern genetics.

At the turn of the twentieth century, "farmers and breeders emphasize the very great importance of good seed"—a philosophy that shaped social policy in at least thirty nations around the world before World War II.[28] Coined by Francis Galton, Darwin's cousin, the word *eugenics* meant "well-born," a term that was rooted in a view of animal breeding. Agricultural genetics, begun with Bakewell, provided the model for what would become eugenics, the idea of studying and manipulating human populations. Biology was viewed as the "basis of a great number of perplexing social problems," wrote Thurman Rice, in *Racial Hygiene: A Practical Discussion of Eugenics and Race Culture*, an influential book published in 1929. "The welfare of the race rather than that of the individual is the aim," Rice insisted. "However, the welfare of the race will ultimately mean the greatest good to the greatest number of individuals."[29]

Rice echoed Progressive ideals about improving the individual and society. Among examples of poor human stock everywhere, the Jukes family became famous for their deficiencies. The Jukes, a large backwoods family in New York state, had been studied in 1877 and again in 1915, and their lack of progress in American society was blamed directly on their poor genetic stock. Of the 2,820 family members studied, many were criminals, paupers, prostitutes, and "defectives." The study emphasized how much the family had cost the state due to crime and disease. There were six hundred "defective" Jukes in 1915—ready to spawn more, the public feared. It became a strong argument for sterilization.[30] The Kallikak family, another study that compared offspring, contained many members who were feeble-minded, prostitutes, and alcoholics. There seemed to be destitute broods in every backwoods, and they were propagating more, at potential public expense. Books like *Racial Hygiene* emphasized the number of these families that were in poor asylums, orphanages, schools for the feeble-minded, reform schools, prisons, and hospitals. "This is where the state money goes in large part," Rice noted. "This is one of the reasons why taxes are so high. These families are extreme . . . but they exist. . . . Each succeeding generation they present a bigger and bigger problem."[31]

Meanwhile, good stock—for example the "Virginia Aristocracy" and "New England stock," Rice noted—were having fewer children. Even Charles Darwin's family and Cleopatra's lineage were detailed in the book. Both included first-cousin marriages, and Rice pointed out that this inbreeding had not harmed the Darwin progeny at all, in fact they were all brilliant. He cited other examples of brilliant offspring from first-cousin marriages, going so far as to speculate that the original New England stock had been so few in number that "there must have been much inbreeding, but without bad effects since the stock was good."[32]

Eugenics did not proceed the same way livestock breeding did, though. While cattle breeders sought to maintain and improve positive traits over the generations, eugenics focused on negative traits, attempting to cull the human populations of their least productive members. Club feet, extra digits on the hand, mental disorders, epilepsy, and eventually finer and finer degrees of "imperfection" were the targets. At the same time eugenicists feared that highly intelligent people would produce smaller families (about two children) while the allegedly degenerate elements of society were having larger families of four to eight children. Public welfare might play a role in allowing these less fit people to survive and reproduce—thus disrupting the normal flow of natural selection that was moving along a path of "progress" toward the fittest. Highly influenced by Darwin's observations of the process of natural selection, eugenics applied the "survival of the fittest" to humankind.

If people ought to be progressing toward perfection, how to explain criminals? Perhaps they had been *born* that way, throwbacks to a primitive stage of human evolution. In the late nineteenth century Dr. Cesare Lombroso, an Italian criminologist, had presented the idea that criminals were humans who had for some reason reverted back to their prehistoric ancestors—earlier, less civilized. Lombroso, a professor of medicine and criminal anthropology at the University of Turin, also developed the idea that a criminal could be identified by physical characteristics. If a criminal was "born," not "made," society could not be blamed for his faults. He had been created that way. This led to more humane treatment of criminals, at least in Europe. It also prompted society to ask how to prevent criminals from being conceived in the first place. If society was to be pure and move toward an ever higher degree of evolution, criminals ought to be eliminated.

Physicians accepted this idea that social failure was a medical problem. They attempted various treatments, but when medical means failed to help the psychotic, the retarded, the pauper, and the vagrant, eugenicists shifted to preventive medicine. Rudolph Virchow, a leading German physician and roughly a contemporary of Lombroso's, led a public health movement which sought to deal with health problems on a grand scale. United with the ideas of eugenics, Virchow's efforts became the racial hygiene movement in Germany, which later traveled to the United States via the physicians who went to German medical schools for training.

How to keep criminals and "defectives" from multiplying was

the main concern. The solution was custody in asylums or compulsory sterilization. Most doctors probably felt that sterilization was more humane. Vasectomy and tubal ligation were employed because they were highly effective yet allowed the person to reenter society rather than remaining institutionalized at public expense. Proponents believed that sterilization was in the best interests of everyone, because social failure was seen as due to the germ plasm, not the individual or his environment.[33]

How that germ plasm was understood in 1900 relied on Mendel's 1865 paper on peas, where he noted that when crossing yellow and green peas, a particular pattern of inheritance could be predicted. Known as Mendel's laws of inheritance, this idea was soon extended beyond garden peas to explain how animals, and even humans, received their inherited traits. By studying family trees, scientists could determine a pattern of recessive, dominant, or sex-linked inheritance. Today DNA markers are used to follow that pattern, but in 1900 the cutting edge of research was found by interviewing people with a particular trait (usually an affliction) and asking about family members in order to create a family tree of the trait spread over generations.

In 1910 the Eugenics Record Office was established at Cold Spring Harbor, New York, as the center for U.S. eugenics research. From there, researchers fanned out across the country, searching for large families that suffered from medical conditions that might be heritable. They visited circuses (where people with unusual attributes often worked as "freaks"), prisons, insane asylums, orphanages, and homes for the blind. The public cost of maintaining such individuals was emphasized by eugenicists as they sought to discern ethnic patterns.

By World War I and the advent of intelligence testing, the concerns grew more serious. Recruits for the military service who were native-born Americans scored significantly higher on IQ tests than immigrants did. This fact contributed to calls to limit immigration after the war's end. Proponents of restricted immigration claimed that immigrants from southern and eastern Europe were entering the United States with high levels of mental illness, crime, and social dependency. Across the country during the 1920s and 1930s, state fairs held Fitter Families exhibits, where fairgoers could receive a eugenics evaluation and perhaps win a medal for their inherited makeup. The Miss America pageant, a promotional venture begun in Atlantic City in 1921, capitalized on the era's obsession with evaluating individuals for physical and mental characteristics.

The "overenthusiastic eugenicist" was the focus of a 1932 article by T. Swann Harding in the *Scientific American* which reminded readers that eugenics was for cows, not people. "Breeding up the human race probably is a charming and certainly a fascinating occupation but one that should be restricted to armchair biologists," Harding warned. "Unfortunately, this is not the case. The assumption that it would be quite as easy to breed humans up to the level of our 'best' people as it is for men to breed cattle with desirable qualities is frequently invoked. This assumption is not only unscientific; it is ridiculous." The idea was impractical, Harding explained, because the eugenicist would need "a stud-farm where he can secure control, isolation, and purity of blood"; desirable subjects would be the least willing to surrender their "liberty" to join the endeavor.[34] Besides, who would control such an effort? And who would decide what attributes should be bred for?

Using the dairy cow as an example, Harding pointed out that it was "not a very successful animal. For ordinary bovine purposes she is really a rather grossly malformed creature—a sort of monstrosity suffering from an hypertrophy of the lacteal glands. This can do her very little good personally. As the process proceeds to perfection (from the dairyman's standpoint), she becomes more and more an animated milk-secreting machine, but purely as a cow she is a far worse bovine than when the process began."[35]

Still, there were many "biological enthusiasts" who believed that animal breeding had been so successful that methods used in agriculture could be replicated in order to breed up the human race. Scientists responded that even cattle breeding was complex, and much was unknown. Besides, cows went through a three-year maturity span that made it possible to determine whether matings had been successful or failures (for increased milk production). As the *Scientific American* noted, "It requires 12 to 25 years to find out whether a human mating has or has not transmitted high intellectual attainments." A long maturation time would hinder efforts to up-breed humans, but more likely the human race would "offer more resistance to compulsory matings than the cow." Ultimately the result was that "biologists of no matter how great prestige who claim we could easily increase the quality and intelligence of the human race by selective breeding are talking practically at random. That they still further mislead the fallacy-ridden human race is a misfortune."[36]

S. J. Holmes, a professor of zoology at the University of California, wrote in a 1936 book, *Human Genetics and Its Social Import*, "The notion that human beings, like domestic animals, may be improved by selective breeding never attained wide currency until recent years. . . ."[37] With a depression-era viewpoint, Holmes noted that people were now concerned about how populations would support themselves. "If romantic love and parental affection were given no weight, and human beings were bred like so many cattle, the race could doubtless be greatly improved in quality in a few generations," Holmes wrote.[38] He used pedigree charts to show how epilepsy, Huntington's chorea, insanity, and other problems were inherited as recessive genes, which emerged when individuals who carried them had children. Holmes even suggested that "many new gene mutations will arise which will result in hereditary defects never heard of before."[39]

In Europe, meanwhile, a unique development kept interest in human breeding alive. Scientists had truly bred backward, using cattle: zoologists in Germany had been able to breed a "throwback" in a frighteningly short time. If that could happen to cattle, might humans be doing it to themselves? The phenomenon of reversion was familiar to animal breeders who crossed two animals of different colors, resulting in offspring of a third color. The crossbreeding had resulted in a "throwback"—an ancestral trait that had been recessive now emerged. This idea, that people could mix and mate and produce children who moved backward on the evolutionary scale was unsettling.

Heinz Heck, director of the Munich Zoo, had dabbled in breeding back to resurrect an extinct species of cattle, the ancient aurochs. Heck wrote, "The last Aurochs, a cow, died in 1627 in a Polish park, and that was the end of one of the finest animals—the powerful and colorful wild ox, an animal to which mankind owes much, since our civilization can hardly be imagined without its most important domesticated animal, the cow."[40]

Zoologists knew what aurochs were like. There had been numerous archaeological finds of skeletal remains, and the cave paintings at Lascaux and elsewhere had shown their coloring. Bulls were described as over six feet tall at the shoulder and usually with very long horns. They were black with a cream-colored stripe down the back and on the forehead. Cows were mostly brownish-red, some with patches of black or cream color. Calves of both sexes were red-brown until they were about six months of age. Very much like the paintings in French and Spanish caves, the aurochs were animals that had existed side by side with stone age man but failed to thrive in the changing Europe and became extinct. European bison, a cousin, continued on in Poland, Lithuania, and Prussia.[41]

By crossing breeds of cattle that displayed characteristics of their wild ancestor, Heck calculated that an animal could be bred to contain many of the same genes as the aurochs. While not identical to the ancestor, it might be quite similar and breed true. "No animal," Heck said, "is utterly exterminated as long as some of its hereditary factors remain." That these qualities were not visible only validated the laws of heredity; recessive genes appeared when properly manipulated through selective breeding. "For what is hidden may be brought to light again," Heck pointed out, "and, by cross-breeding, the original component parts may again be isolated."[42] Heck could dip into those ancient genes that were present in a variety of different breeds of cattle, using one that had an aurochs's horns, another its build, another its color.

Remarkably, Heck achieved back-breeding "incredibly quickly." He crossed "all kinds of races of cattle in a way that would have horrified a pedigree breeder. Hungarian and Podolian steppe cattle and Scottish Highland cattle were mated with grey and brown Apline breeds from Algau and Werdenfels and with piebald Friesians and Corsicans." He even bought a few cross-breeds, and "all were thrown into the pot so to speak," he said. By the spring of 1932 he had achieved the "first good specimens of the Aurochs of modern times, one of each sex." Heck realized he had achieved a landmark in livestock breeding: "It was like a miracle. The first Aurochs for 300 years could be seen alive."[43]

To his astonishment, succeeding generations did not lose their characteristics, and there was not "one throw-back to any of the domestic breeds used. The calves are all as alike as slices of bread from one loaf," he wrote in 1951. His brother, Professor Lutz Heck, tried the same experiments in the Berlin Zoological Gardens, using different cattle. Lutz cross-bred Spanish and French fighting breeds with other Mediterranean breeds, and the results were identical—the aurochs created at Berlin were exactly like those at Munich. The Heck aurochs looked exactly like their earlier relatives—the French and Spanish cave paintings. The horns looked like fossil aurochs horns.[44] Not only did these reconstituted aurochs look like their ancestors, they *thought* like them too. They were fierce, temperamental, and quite fast and agile. They could turn shy and hide in the woods of their compound.

The fact that the aurochs was created twice, and separately, proved that cross-breeding was powerful. The "wild" was very close to the present. Whether or not the Hecks' cattle experiments were designed to show the consequences of "race mixing," the results were chilling to those proponents of selective breeding who yearned to save particular races from "race suicide." Were Germans shocked at how quickly a breed could return to primitive status? Could humans become primitives in only a few generations? Mixing with other, "lesser-accomplished" races might do what the reconstituted aurochs revealed: one's mental as well as physical makeup might revert to the primitive.

The German eugenics movement began as a humanitarian effort to improve the population and was supported by the educated middle classes; it was essentially the same as the eugenics movement in the United States. But eugenics and genetics were fields in confusion in the early twentieth century; legitimate scientific studies were joined by amateurish work that bordered on pseudoscience. Medicine had ignored, even avoided any research in human genetics. Human health care focused on malnutrition and fighting infection—both environmental factors. Eugenics was too abstract, too theoretical.[45] Had physicians become involved, the direction of eugenics might have been shaped along more objective, more scientific lines. But the disinterest of the medical community, along with confusion and lack of consensus in the field, prevented the study of human genetics from achieving a scientific standing and allowed eugenics to be swept up in political issues.[46]

Amateur eugenicists, such as W. E. D. Stokes, a horse breeder, moved to the forefront of the movement. Stokes claimed, "There is no trouble to breed any kind of man you like, 4 feet men or 7 feet men—or, for instance, all to weigh 60 or 400 pounds, just as we breed horses"—this in a 1917 issue of the *Eugenical News*.[47] Racists too were heavily involved in the movement. "Strong in political sentiment, lacking in scientific interest, racially and culturally prejudiced," Kenneth Ludmerer writes, "these men found in the movement a scientific sanctuary."[48] As the world recovered from World War I, the eugenics movement, seemingly based on sound science and sociology, increased in popularity and esteem. The next step was legislation.

In 1917 sixteen states enacted compulsory sterilization laws for certain "unfit" categories of people. But the laws were overturned or ignored. Soon, however, the situation changed. By 1931, fresh from the Crash of 1929, thirty states had passed sterilization laws. It was during the depression that American eugenicists helped Germany formulate the policy that Hitler would adopt in 1933.[49]

Along with sterilization, race now became an American issue. The theory of white superiority over nonwhites had survived the Civil War and in the 1930s was refined to include finer definitions of the white race. Immigrants from southern Europe, the Baltics, and Ireland were often disparaged as ethnics who might displace Anglo-Saxon populations in the United States. Economics was certainly at the root of the issue: in hard times, jobs and financial security were threatened by immigrants.

Adolf Hitler based his racial ideology on the idea of Aryan superiority as a means for national regeneration. His ideas were palatable to Germans because their country had been so scarred by defeat in World War I and the burden of reparations followed by depression. "Rassenhygiene," the German race hygiene movement, was supported by the educated middle class in an era when Caucasians were accepted as a superior race. Germany was a homogeneous society, and what we term racism today was fairly common thinking in the early twentieth century. The German race hygiene movement emphasized increasing the number of "better" citizens while eliminating the numbers of "unfit" who were a drag on the economy and the cultural progress of the nation. Fitness equaled achievement in Germany; degeneracy was evidenced by anti-social behavior and the inability to contribute to society. But Germany was by no means alone. Before Hitler's rise to power in 1933, German eugenics had been little different from eugenics movements in other Western countries.[50]

The historian Sheila Faith Weiss gives several reasons for the rise of eugenics in Germany, including the shift to industrial urbanization which created labor unrest; the rise in social problems such as crime, alcoholism, insanity, and suicide; and the adoption of eugenics by medical professionals as preventive medicine. Eugenics was seen as one way to improve the general level of public health without great financial outlays. Germany's economic plight in 1929 had led many to question whether the country could continue expanding the welfare state. Productivity was vital, everything else had to be eliminated. Asylums, hospitals, jails—all were needless expenses caused by unfit individuals. "Thus, people became a manipulable resource," Weiss writes, "to be administered in the interest of a healthy and culturally productive nation."[51] Germany's new planned economy would have to include eugenics as part of its health policy.

The United States had been sterilizing "defectives" since 1907. While German eugenicists at first avoided advocating voluntary sterilization for hereditary defects because they knew the public would oppose it, as time passed it began to look like a solution. In 1932 a law permitted voluntary sterilization if there was proof that the individual carried a hereditary defective train. Only the Catholic church protested. In 1933, after the Nazis came to power, that law became the basis for Germany's sterilization law.

Race hygiene now became the key to Adolf Hitler's vision for Germany. Nordic blood and preservation of the Aryan race were Hitler's obsessions and central to Nazi political goals. What had begun as a rational scientific approach to social problems quickly evolved into such Nazi race policies as euthanasia, extermination of Europe's Gypsy population, and the "final solution" for the Jews. Other countries had enacted sterilization laws—the United States, Finland, Sweden, Norway, Iceland, and Canada—but Germany's legislation moved quickly. During its first year (1933), more than fifty thousand "defective" Germans were sterilized. In 1939 Hitler legalized euthanasia, and within two years fifty thousand people had been put to death.

U.S. eugenicists had not foreseen how their principles might be corrupted by the Nazis. In 1933, as the Nazi effort commenced, the *Eugenical News* noted that "It is difficult to see how the new German Sterilization Law could, as some have suggested, be deflected from its purely eugenical purpose, and be made an 'instrument of tyranny' for the sterilization of non-Nordic races."[52] The Nazi "breeding program" aimed to eliminate any public cost related to supporting the unfit, and eventually to eliminate them. At the same time the government encouraged "fit" individuals to have large families. Under mandatory sterilization laws, over five years from 1934 to 1939, between 200,000 and 400,000 feeble-minded, mentally ill, blind, epileptic, or alcoholic Germans were sterilized.[53]

It is impossible now to assess the extent of Nazi eugenics research during World War II, as so many records were destroyed. Racial cross-breeding was tried, but there were no attempts at mixing Aryans and Jews—that was illegal. Citizens of questionable background had to obtain documentation of racial ancestry, and some children of mixed races were sterilized. As the movement progressed, euthanasia was adopted as an economical solution. Between 1939 and 1941 about 100,000 "useless eaters"—mentally ill or retarded people—were euthanized.

By the time the German program was in full swing, eugenics in other countries had spent itself. The depression undercut British and American notions that the underclass was a product of heredity. The number of poor had increased faster than their birthrate, and individuals from good bloodlines were as poor as anyone else. In Russia, Bolshevik plans for centralized planning would have made eugenics easily enforceable, and the government did flirt with it. The idea of planned artificial insemination, patterned after cattle breeding, interested Aleksandr Serebrovsky, a leading Marxist biologist, who wrote in 1929 that "One talented and valuable producer could have up to one thousand children. . . . In these conditions, human selection would make gigantic leaps forward. And various women and whole communities would then be proud . . . of their successes and achievements in this undoubtedly most astonishing field—the production of new forms of human beings."[54] Serebrovsky had been an agricultural livestock and poultry breeder and had spent years traveling to remote regions looking for chickens with unique genes to breed up desirable characteristics. During the 1920s Russians had adopted the widespread use of artificial insemination in cattle, even sending researchers to West Africa to attempt a hybridized ape by artificially inseminating chimpanzees.[55]

But Stalin had other ideas. He viewed artificial attempts to shape the population as too bourgeois, too reliant on "experts." He distrusted ideas that treated people like resources to be manipulated and managed for profit. By the 1930s in the USSR, eugenics was dead—at least the model created by animal geneticists like Serebrovsky. In spite of the lure of creating a refined population, which would have dovetailed with the Soviets' forced collectivization of agriculture, Stalin disliked the idea of artificial insemination. Several of its proponents were shot in the late 1930s, and the idea of medical genetics dissolved.[56]

The eugenics movement was not a European-American idea.

Brazil, Mexico, Argentina, in all thirty countries experienced national eugenics movements. And it was not founded on ideas of race and class but on the use of science in a socially responsible way. It was never left- or right-wing but was supported by both sides. Many women were involved—Margaret Sanger, for example, who led the birth-control movement in the United States. Many leading eugenicists were Jewish, in fact many eugenics texts use Jews as examples of ability and achievement as a result of generations of strong selection. Jewish eugenicists had been involved in the German movement until the Nazis replaced them with Aryan scientists in the mid-1930s. Countries developed eugenics values that grew out of their own nationalistic goals. In Mexico, for example, race mixing was promoted as a way to produce an improved national population.[57]

Eugenics had been a dynamic answer to social problems. It allowed people to interpret "science" on their own terms and embrace it as a way to create a better future. It came out of the era of Progressive efficiency—Taylorism, Ford's production lines, scientific rationalism—at a time when diminished economies pointed to a Malthusian future. Americans had been moving from rural to urban areas, and significant numbers of immigrants had come to live in industrial regions. For many, educational opportunities were nonexistent.

Based on livestock breeding, eugenics simply was not tenable in large human populations. While livestock breeders continued their experimental programs, defining pedigree and embracing artificial insemination as a way to breed even more closely on ever larger numbers, people were not willing to continue. The price was too high. Who would, or could, determine which human characteristics should be maintained and which eliminated from the genetic population?

DARWIN HAD STUDIED the effects of cross-breeding in plants and domestic animals and was interested in how breeding two different family lines might create a reversion to ancestral types.[58] The animal breeders of his day were more interested in selection than in cross-breeding, which helped shape his own ideas about natural selection. Some breeds developed on their own, due to genetic mutation and isolation from other animals to cross-breed with. The environment itself can create breeds, eliminating those who do not hold certain traits—for instance, the N'dama cattle from West Africa, which over centuries have developed a resistance to trypanosomiasis, or sleeping sickness, spread by the tsetse fly, which is fatal to other cattle. Texas longhorn cattle that had reverted to the wild developed longer legs and greater agility as well as a resistance to tick-borne diseases. Other stock that was short-legged and susceptible to disease died out. Heinz Heck noted that the "Auroxen" he created were "almost immune to cattle diseases that cause havoc among domesticated herds, namely foot-and-mouth and rheumatic fever." Heck's cattle had been infected with those diseases but became only mildly ill, then were immune. At about the same time, Germany experienced an epidemic of cattle diseases that wiped out its dairy herds.[59]

The value of particular breeds of cattle changes over time. For example, the Dutch Belted cow, which was an excellent dairy cow on grass-based farms in the early 1900s, has fallen from favor, replaced by the Holstein, which yields more milk from a cereal grain diet. If cereal grains were no longer economical as dairy cow feed, perhaps the Dutch Belted would again be preferred. If the environment were to change—for instance, if the overuse of antibiotics resulted in their complete abandonment in fighting cattle disease—other strains of cattle that have inborn resistance to particular diseases, such as the Scottish Highlander breed, would be highly desirable. Maintaining the genetic diversity of the cattle species is imperative for the future, both for cattle and the humans who depend on them.

Heck's reconstituted auroch in Berlin were lost during the war, but those in Munich continued to be bred, and at one time there were more than fifty of the "Heck cattle" in Europe. Today there has been a significant revival of interest in the reconstituted aurochs, and a multi-national organization known as SIERDAH, the

Syndicat International pour l'Elévage, la Réintroduction, et le Développement de l'Aurochs de Heck, has been formed in France. Zoos throughout France, Belgium, Switzerland, the Netherlands, Hungary, Sweden, Spain, and England have joined the organization, and private farms are now raising aurochs in Europe. Efforts are directed at studying the animals for their potential because they can live in a wide range of environments and appear to be disease resistant. Environmental efforts by the World Wildlife Fund in Holland to reintroduce cattle into natural environments have concentrated on bringing back the original European grazing bovine, Heck cattle, as a way to restore the natural landscape.[60]

From his vantage point in postwar Germany in 1951, Heck claimed to have carried out his experiment because of "the thought that if man cannot be halted in his mad rage for destruction of himself and all other creatures, it is at least a consolation if some of those kinds of animals he has already exterminated can be brought to life again."

Heck's aurochs linked us to a primeval past, yet it also proved to be a match that lit the flames of genocide. Today, as we stand on the precipice of genetic manipulation beyond anyone's wildest dreams, it is well to remember that what science creates may be interpreted and applied quite differently by society.

12

A Shot in the Arm

They do not believe that it affords an efficient and assured protection against the invasion of the small pox; they have a natural disgust to the idea of transferring to the veins of their children a loathsome virus derived from the blood of a diseased brute and transmitted through they know not how many unhealthy human mediums; they have a dread, a conviction, that other filthy diseases tending to embitter and shorten life, are frequently transmitted through and by the vaccine virus. . . .
—John Gibbs, *Compulsory Vaccination*, 1856

. . . nonsense about the danger of inoculating humours from an animal whose milk makes the principal part of our children's food, whose flesh is the source of Old English courage, and whose breath is not only fragrant but salubrious.
—Joseph Adams, *Answers to all the Objections Hitherto Made Against Cow-Pox*, 1805

EVEN TODAY, if told we could protect ourselves from a deadly infectious disease only by introducing matter from someone else's swollen, infected pustules into our own body, it would be difficult to accept. Even more distasteful is the idea of inserting pus matter from a sick animal into the human body. In 1798, as the science-crazed eighteenth century tapered to a close, Edward Jenner, a forty-nine-year-old English country doctor, self-published his first book, *An Inquiry into the Causes and Effects of the Variolae Vaccinae.* That book not only made him a national hero and center of worldwide controversy, it unleashed the most unlikely breakthrough in the history of medicine: human vaccination with animal lymph.

Jenner's discovery that matter taken from an infected animal—a cow—could give a person protection from the most dreaded disease of the day, smallpox, was highly controversial. Earlier in the eighteenth century, *variolation* or *inoculation* against smallpox had been discovered. It involved applying human pus matter to an opening in the skin, which caused the body to develop immunity against smallpox.

The idea of introducing infected matter into a cut in the skin had been a difficult one to accept in the 1720s when Cotton Mather, a prominent Boston minister and scholar, first learned of the procedure from his African slave, who had been variolated years before. Mather was fascinated by the way this "grafting" (a term adopted from the plant-propagating experiments of the day, which involved grafting stems onto root stock) gave lifelong protection against the dreaded disease smallpox. He wrote letters and articles in an attempt to introduce it to England's academic and medical elite, realizing that if the British accepted it, Americans might view it in a more favorable light. But he was ignored by the English medical and scientific world, and Bostonians were not about to accept the medical teachings of African slaves, no matter how successful their treatments might be.

Mather's efforts were stymied by a medical community that viewed inoculation as a threat to their own livelihood. Treating smallpox victims was, after all, profitable business. Rallying the public through the press, the doctors and their supporters (several were tobacconists and pharmacists who profited from selling smallpox tonics) filled local newspapers with inflammatory accusations. The *New England Courant*, owned by James Franklin, older brother and employer of young Benjamin Franklin, established itself as the newspaper of the anti-inoculators. The public was sufficiently stirred to riot in the streets. Mather and Dr. Zabdiel Boylston, the one physician who performed inoculations, were scorned by the general public as well as the medical community. Boylston and his family went into hiding; Mather's home was firebombed by an angry mob. The clergy, under Mather's leadership, promoted inoculation as a public health measure, but the medical community continued to fight it for reasons of business and fear that it would spread disease.

Inoculation involved pricking the skin and introducing matter taken from a smallpox victim's pustules into the opening. The patient soon came down with smallpox, but a far milder case than if the disease was ingested into the lungs, the natural mode of infection. Inoculation incurred immunity against smallpox for the rest of one's lifetime. It was an ancient technique, used in southern Asia for centuries before it entered Africa, and came to the North American colonies with slaves who had learned of it in their homeland.

Anyone could inoculate if they had access to scabs or pus from true smallpox patients. The term "buying the disease" came to mean inoculation. One could buy the disease from sources who sold scabs or threads impregnated with dried pus. Once moistened, the thread could be tied against a cut in the skin, and the job was done. Threads were sent through the mails, even sent from Britain to America and around Europe. And inoculation was something anyone could do—quacks, herbalists, women. It could not be limited to the medical profession. If anyone could obtain smallpox matter, they were in business.

The major drawback of inoculation was that patients felt so good they often went out and infected others, engendering epidemics of the disease. They were harboring a case of true smallpox, and anyone naturally infected by them came down with the virulent form of the disease, from which many died. Inoculators were often irresponsible about quarantining their patients until they could no longer spread the disease, and the patients often did not understand the danger they presented to the community. Experience had shown that when inoculators moved to a town, smallpox epidemics often ensued.

Smallpox is believed to have originated in India or China, where it existed for thousands of years before it was described in Greek or Roman texts. It moved into Europe as a result of the Moorish conquest of Spain and was taken north by Crusaders returning from the Holy Land. It was not present in the Americas until the Spanish introduced it, notably by one soldier who carried the infection with Cortés's army into Tenochtitlan. That lone case is believed to have caused an epidemic which wiped out the Aztecs and allowed Cortés to take control of what is now Mexico. As Europeans moved across the American continent, smallpox swept away millions of Native Americans, just as it had done earlier in Europe. During the seventeenth and eighteenth centuries, smallpox had wiped out between 200,000 and 600,000 Europeans each year.

The disease affected everyone, elites and commoners; the mortality rate after infection from smallpox was between 12 and 25 percent, but in many areas it reached 50 percent or higher. It accounted for one-third of childhood fatalities. Survivors were often left badly disfigured or blind. There was no cure for it, though it provided a thriving market for quackery and tonics for the desperate. Inoculation, once tried and proven, was the only protection.

But inoculation cost money; the several weeks of preparation it required (purging, bleeding, and a restricted diet were thought imperative for success) as well as extended bed rest and seclusion in a hospital put inoculation beyond the reach of the working classes. If workers were inoculated and allowed to return to work, they would engender epidemics among those not yet inoculated. The London Smallpox Hospital, founded in 1746, provided free inoculations, but who could miss work? The hospital eventually served only the wealthy as a place to send their house servants for inoculation. And unless an epidemic threatened or actually broke out, few people bothered to go through the uncomfortable and slightly risky procedure of inoculation. The time was ripe for a better alternative.

Edward Jenner was a difficult, self-centered, uncooperative prima donna. But he changed the face of scientific medicine when he discovered the use of cattle lymph in inoculation. Some biographers portray Jenner as a genius, others attempt to expose him as a rogue. He was born in 1749, the third son of the Reverend Stephen Jenner of Berkeley, Gloucestershire. As a boy he was fascinated with natural history, as most educated people were at the time. He was a trained physician—at least he had finished an apprenticeship and received his medical degree from St. Andrews University in Scotland, *in absentia*. Most medical degrees at the time were similarly awarded; two physicians wrote letters of recommendation for Jenner, and the school sent the degree.[1]

After Jenner married he moved to Cheltenham Spa, a fashionable health resort, where he set up a medical practice for the gentry. Eager to establish credibility for his elite clientele, he sent articles to the Royal Society about his various experiments. He reported the mating of a dog and a fox, and wrote about his trials using fresh human blood on growing plants. He found manure to be "destructive to vegitable life"[2] and figured the stomach was the "governor of the whole machine, the mind as well as the body." Much of his time was spent dissecting dogs.[3] Eventually his article about the cuckoo bird was accepted for publication in the Society's journal, *Philosophical Transactions*, and he received membership, but the article was of dubious value. Jenner later admitted that he had invented conclusions about how the cuckoo bird was able to inhabit the nests of other birds.

How Edward Jenner got the idea to take cowpox matter from a dairy cow and inoculate it into the skin of patients as a protection against smallpox has never been very clear. He lived in a rural area where livestock were handy for experimentation and observation. And it was common knowledge in rural dairying areas that anyone who milked cows and came down with pox from the cow, called cowpox, or kinepox, would break out with pustules on the hands that later healed. Afterward, for whatever reason, those persons were resistant to infection from smallpox—an entirely different disease—for the rest of their life. So commonly held was this knowledge that milkmaids were frequently hired to care for families when smallpox epidemics passed through, because the milkmaids were known to be immune to the sickness. Edward Jenner, casting about for a medical mystery to sink his teeth into, gave up his dog autopsies and blood-fertilizer experiments to investigate the connection between cowpox and smallpox.

In 1789 there had been an outbreak of swinepox among pigs in Jenner's community, and his infant son's nurse had come down with the infection. He inoculated the baby and two of his neighbor's servants with matter taken from the nurse's pustules. He may have been trying to protect them from smallpox or simply experimenting, as he did with his dog and blood experiments. Jenner had been inoculated with smallpox matter as a young boy, spending weeks in a "smallpox stable" with several other boys of the upper class, so he was immune to the disease and unable to experiment on himself. There was no smallpox active in the area at the time, but a year later, when smallpox appeared, he obtained matter and inoculated the servant and his son again, checking to see if the swinepox inoculation had immunized them against smallpox. They had no reaction; the swinepox inoculation had protected them. But Jenner now dropped his work with swinepox, perhaps because it died out in the area and he was unable to obtain more matter. His wife, Katherine, supported his scientific efforts, and a few years later she allowed Jenner to inoculate their second son with matter from a diseased horse—but the results were negligible.[4]

When cowpox appeared in the local animals a year later, in 1796, Jenner resumed his research. He found Sarah Nelmes, a milkmaid who had become infected with cowpox, took matter from the pustule on her hand, and inoculated a young boy, James Phipps, with it. Later, when he was able to obtain virulent smallpox matter, Jenner inoculated the boy with it. He did not come down with the disease; the cowpox inoculation had protected him.

Elated, Jenner wrote to a friend, ". . . Now listen to the most delightful part of my story. The Boy has since been inoculated for the Smallpox which as I ventured to predict produced no effect. I shall now pursue my Experiment with redoubled ardor."[5]

Jenner had not stumbled onto anything remarkable, at least in the eyes of rural English folk, who had known for nearly a century that a case of cowpox made one immune to smallpox. It was common knowledge in cattle districts, and after Jenner's first paper about cowpox inoculation was published, others claimed to have preceded him. Dr. Fewster had read a paper to the London Medical Society titled "Cowpox and Its Ability to Prevent Smallpox," but it had made no impression on the public. Others had inoculated cowpox (or claimed to) in Germany in 1769, and in Holstein and Scandinavia in 1791.[6] Benjamin Jesty, a Dorsetshire farmer, claimed to have inoculated his wife and children with cowpox in 1774. When attention focused on Jenner, critics brought up Jesty's story and honored him with a ceremony at the Original Vaccine Pock Institute in London.

Clearly there were others who had been using the technique before Jenner, but it was his publication of the *Inquiry*, in 1798, that brought the procedure from the shadows of folk medicine to the awareness of the literate. Now it could be tried, tested, and dispersed.[7] Others may have done it, but Jenner was the one who applied scientific method, who went on to *prove* that inoculation could be done effectively person-to-person, not only directly from cattle.

Yet Jenner did not think he was working with a cattle disease; he insisted that he was investigating a disease of horses which he believed had infected cattle in the area. Confusion thus surrounded Jenner's work, and to this day it's not certain exactly what he was working with, or if he had actually crossed cowpox with another virus, perhaps smallpox or vaccinia, another version of smallpox virus. Cowpox was not common, and he believed it resulted from men who worked with horses passing a horse infection known as "grease" to dairy cows. In eighteenth-century England, men worked with horses, the draft power for cultivation and transport, while women worked with dairy cattle. Men did the milking only in emergencies; otherwise dairy production was solely in the hands of women. This sexual division of labor with regard to livestock prevailed in Scotland, Ireland, and Scandinavia too. Jenner thought perhaps men infrequently passed horse grease on to cows.

When the Royal Society refused to publish Jenner's book, he paid to publish it himself. *Inquiry* proved to be a turning point in medicine; it marked the first efforts to introduce animal cells, in this case from cattle, into the human body in order to prevent disease. The monumental discovery that infected matter from cows could be inserted in the human body and prevent infection from the most dreaded disease of all time could not be ignored. Critics emerged immediately: the idea was inane, even obscene. Cartoons appeared showing people's bodies growing cattle parts. Ridicule and fear characterized the debate that raged as Jenner's idea spread. After all, cowpox was a sexually transmitted disease of cattle—a sort of cow syphilis. Who would knowingly infect themselves with such a thing?

One description of Jenner, appearing in a pamphlet in 1807, announced that "a mighty and horrible monster, with the horns of a bull, the hind hoofs of a horse, the jaws of the kracken, the teeth and claws of a tiger, the tail of a cow,—all the evils of Pandora's box in his belly,—plague, pestilence, leprosy, purple blotches, fetid ulcers, and filthy sores, covering his body,—and an atmosphere of accumulated disease, pain, and death around him, has made his appearance in the world, and devours mankind,—especially poor, helpless infants. . . ."[8]

Ultimately his countrymen embraced Jenner as a symbol of British genius. The English had lost their American colonies and needed a national hero when the discovery of vaccination turned the tide for English science. Jenner had stumbled upon an incredible idea that no one else would likely have thought of for centuries. He had been open-minded and curious enough to persist with his experiments and had opened a scientific door for others. He could have hoarded his knowledge, kept the technique a secret for himself and his children, just as many inoculators had done with procedures they had devised. But this discovery brought Jenner what he sought: attention and respect from society. The support of colleagues, the admiring fans, the deluge of mail he diligently answered were his rewards.

While Jenner has been hailed as a godsend, Dr. Derrick Baxby, senior lecturer in microbiology at the University of Liverpool, has noted that "even now we do not know where vaccinia virus came from."[9] Analysis of smallpox, vaccinia, and cowpox viruses shows them to be completely distinct from one another yet all members of the same family, *orthopoxvirus*. Vaccinia is the virus now used in smallpox vaccination, and oddly it does not exist in nature and may not have existed before Jenner began using cowpox as a vaccine against smallpox. Vaccinia may have mutated from the cowpox virus at some time in the intervening two hundred years, but no one knows when or how it changed.[10] In 1939, when scientific knowledge allowed researchers to examine the vaccinia vaccine they were using, they discovered that it was *not* cowpox at all. They did not know what it was. Or when it had changed.

Vaccination with cowpox matter was too good to be true: the patient did not get sick, and no epidemics ensued. Yet a few years after his discovery, children whom Jenner had vaccinated became sick when smallpox came to town again. The doctor refused to believe that his vaccine did not work, but the event perplexed him. His near-overnight celebrity, the detractors and competitors who profited from his discovery, and the criticisms that the vaccine no longer worked left their mark. By 1810 Jenner's nervous breakdown is evident in his letters, interlaced with odd phrases and statements that do not seem coherent. He realized he was losing his grip on sanity. Growing numbers of children he had vaccinated were coming down with smallpox. Eventually physicians would realize that *two* immunizations were necessary for long-term immunity.

Another concern was the possibility of passing other diseases on to the patient in the vaccine. Of course this had been seen frequently when inoculation had used smallpox matter taken from individuals: syphilis, hepatitis, and erysipelas had all been passed on in the matter. For that reason, inoculators preferred to obtain infectious smallpox matter from children, who might not have contracted other diseases, particularly syphilis. What diseases did animals have that they might pass to humans? No one was sure.

Thus without the threat of an epidemic, the general public was reluctant to pursue vaccination. Complacency led to a massive smallpox epidemic in England in 1840, after vaccination had been available for more than a generation. The Vaccination Act of 1840 provided the first free medical care in England. Still, people were reluctant. It was common knowledge that syphilis had become a part of vaccination by 1814; no one wanted to take that risk.[11] Smallpox might not come around again for twenty years. Meanwhile a sure case of syphilis meant miserable death, before the next smallpox epidemic ever arrived.

One Italian infant whose pustule was used as a vaccine source was found to have passed syphilis to forty-four other infants, who in turn infected their mothers and nurses as well. Fortunately an Italian physician, Dr. Adelchi Negri, solved the problems of transmitting syphilis in arm-to-arm vaccinations. He perfected the technique for propagating cowpox directly in the cow, and passing it from cow to cow. He scarified the cow's bellies before inoculating them, which also created a much larger source of vaccine supply than one person's scattered pustules. Propagating the virus in cows was more reliable and safer. The method spread through Europe to France in 1864 and eventually to England in 1881. Once a source of cowpox lymph was obtained, it could be kept alive in the bodies of other cattle who were inoculated with the matter. Calves were preferred because they were smaller and easier to handle in the laboratory and the hospital. Vaccine matter could be stored in glass vials and shipped elsewhere, and English physicians sent it throughout Britain and Europe as well as to North America. By 1867 inoculated cows were led door to door in Alabama, and children were vaccinated with virus taken directly from the cow at their doorstep.[12]

THE ORTHOPOX VIRUSES are a group found in both animals and humans, and they include smallpox, cowpox, vaccinia, monkeypox, camelpox, and buffalopox. Infection with one will give immunity to the others. Chickenpox is of a different family, the herpes viruses; syphilis, while known as the grand pox, is caused by a spirochete, not a virus. Cowpox was found only in Britain and Western Europe, and until recently was believed to be found naturally only in cows. But cowpox in cattle is rare, and people had contracted it without any contact with a cow. Now it is believed that both cattle and humans became infected when the virus moved from its reservoir in some small, wild animal (as yet unknown). Cowpox was so rare in Jenner's day that he and early vaccinators had trouble obtaining matter from infected cattle.[13] Horsepox, which may have been the source of the grease disease Jenner thought he was using, would have protected humans against smallpox too, but it disappeared in the early twentieth century and is now extinct.[14]

In contrast to earlier difficulties related to the acceptance of inoculation, vaccination came at a time when royal patronage was no longer necessary for the success of an idea; several political revolutions, as well as a new way of thinking, had made acceptance easier. Even the pope exhorted vaccination as a godsend, "a precious discovery which ought to be a new motive for human gratitude to Omnipotence."[15] Clergy in Germany, Switzerland, and England themselves vaccinated people.

A World Health Organization effort to eliminate smallpox through an intensive vaccination program eventually eradicated the disease. By 1971 the world incidence of smallpox was so minimal that continued vaccination was a greater danger. The United States and Britain quit recommending smallpox vaccination for children.[16] The worldwide effort was a huge success but not without its pitfalls. Critics today question what happened when laboratory cultivation of the vaccine moved from the use of calves to the culturing of microbes in monkey kidneys. It has been suggested that the use of the monkey kidneys became the source for infectious AIDS, which was transmitted through polio inoculations prepared on African green monkey kidney cells; the vaccine was used in 1957 and 1960 in Zaire, Rwanda, and Burundi. Human monkeypox virus cases in Africa increased from zero in 1970 to 214 in 1984, in Zaire alone. But a scientific panel tested samples of the vaccine and rejected the idea.[17] Testing showed that genetically it was not close enough to smallpox to generate an outbreak, and infection with the monkeypox virus would act as cowpox did, providing protection against smallpox.

ZOONOSES is the term for animal diseases that are transmissable to people, and they include many of our most common ailments. The interplay between human health and the cattle that humans depend on has gone on for millennia. The medical historian Roy Porter points out that our worst diseases came from proximity to animals. "Cattle provided the pathogen pool with tuberculosis and viral poxes like smallpox. Pigs and ducks gave humans their influenzas, while horses brought rhinoviruses and hence the common cold. Measles, which still kills a million children a year, is the result of rinderpest (canine distemper) jumping between dogs or cattle and humans." According to Porter, smallpox is the result of the long evolutionary adaptation of cowpox to humans.[18]

But Jenner's work shows that animals can be extremely beneficial to humans. The proximity of people to dairy cattle in eighteenth-century England protected them against smallpox and other animal-borne poxes. People who were in close contact with livestock were actually healthier, having received low-level infections in childhood that gave them lifelong antibodies with which to resist later infection. "Herd immunity"—the way animals living close together pass slight infections back and forth and create antibodies against virulent infections—exists in humans too. Polio immunity was discovered to be passed this way. Immunized children shed weakened polio pathogens for up to three weeks after their vaccination, and those they came in close contact with had a chance of contracting the disease; but they were exposed to a non-virulent virus and so became immunized too. Their bodies built up antibodies against polio, as the immunized children did.[19]

We know that the era of epidemics arose when people began settling in high-density cities, where they lived in close, often crowded, proximity to one another. But we do not know whether the fact they had distanced themselves from their domestic livestock was partly to blame. Being no longer in contact with domestic animals (in Rome it was illegal even to have carts and wagons in the streets during the day), people no longer gained immunities from them. Without those immunities gained in childhood, people were highly susceptible when waves of diseases swept through urban areas. As Jared Diamond notes, "When the human population became sufficiently large and concentrated, we reached the stage in our history at which we could at last evolve and sustain crowd diseases confined to our own species."[20] But those crowd diseases had long been present in animal herds. Social animals that lived in groups had long suffered a variety of infectious agents. Viruses are thought to have originated in plants and somehow crossed the species line to infect mammals. When humans began living alongside cattle, pigs, and sheep, their diseases became our diseases.

Cattle keepers have always been close to their animals—in the Spanish Pyrenees the farmhouse sheltered the cattle on the ground floor, the people on the floor above. Other cultures lived in similarly close relationships with their livestock. We have milked them, cleaned up their urine, treated their sores, and drawn their blood. We have worn their skins and shaped eating utensils from their bones. Their infections have become ours. We have always lived in close proximity to cattle dung, a part of the picture too. Dried dung was prized for many uses, particularly as a cooking fuel, a building material in arid climates, and a fertilizer for crops and gardens.

Zooprophylaxis, the study of animals that protect human health, is now a growing field which promises to shine new light on how and why people who live close to domestic animals are often immune to particular diseases. Certainly it's well known that animals have been the reservoir for many of humankind's infectious diseases. But little has been done to show how close contact with livestock, especially cattle, provides a level of immunity against certain diseases. Some researchers now believe that we need a certain level of parasites living within our bodies. Low levels cause the body to create antibodies, which provide protection against larger infestations.[21] Mosquitoes too have been found to create small "vaccinations" by biting people, resulting in antibody formation and immunity to malaria, at least among those who survive such childhood bites and exposure.[22]

Cattle herds have been suggested as the reason why malaria disappeared from the wet, boggy, environment of the British Isles. Malaria once troubled people there, but as the numbers of cattle increased, the threat to humans disappeared.[23] That development had its roots in the Industrial Revolution and subsequent agricultural improvements that resulted in a one-third increase in agricultural productivity. Turnips, like those Bakewell so earnestly grew for his cattle, as well as alfalfa and clover replaced fallow fields and provided feed for cattle on a scale never before seen in European agriculture. Along with enclosure and breeding, these new feeds enabled British cattle herds to multiply to unprecedented numbers. More cattle meant better diets for humans, with increased amounts of meat and milk available at better prices. But another benefit was that cattle provided malaria-carrying anopheles mosquitoes with a preferred bovine source of blood. The malarial plasmodium does not flourish in cattle blood though—it is a dead-end for parasitic infection. With mosquitoes feeding on cattle rather than humans, malaria was interrupted and eventually ended in the British Isles. Not so in warmer Mediterranean countries where few cattle (in proportion to humans) were raised.[24]

Of course it is hard to point to any one cause for the elimination of malaria; enclosure also segregated cattle and sheep into smaller populations and prevented single animals from passing disease between adjacent villages. Along with better feed, the health improvement of the herds reduced the incidence of animal diseases that might be passed to the people, such as bovine tuberculosis and brucellosis. Ultimately the situation was profitable for the British because healthy people worked longer hours and were not sick as often. In France, where enclosure did not occur until much later, the health of the peasants was miserable; malaria and tuberculosis remained a major health problem.[25]

Today, with the development of antibiotic-resistant malaria strains, research is turning to zooprophylaxis, and cattle are being studied as measures for the prevention and control of malaria in Indonesia, Brazil, and even in Afghan refugee camps in Pakistan.[26] Eventually we may come to reevaluate cattle's place in the health of people, particularly those who live in wetlands environments. As northern temperate climates experience global warming, malaria-carrying vector mosquitoes will become a problem in new areas, and cattle grazing nearby may be the most environmentally sound solution—far better for all than toxic chemical pesticides.

Perhaps even smallpox immunity comes from living closely with livestock. Thomas Sydenham, the most accomplished physician of the seventeenth century, wondered why so few "common people die of this disease compared with the numbers that perish by it among the rich?"[27] Many historians think the rich died because of the treatments they could afford, such as purges and bloodletting; others suggest the poor simply died unnoticed. Perhaps, instead, the poor were in closer contact with sources that provided them with immunization as children: on farms they worked closely with cattle, in the towns and cities they trod the manure-caked streets and used the same water troughs as oxen and horses passing through.

Cattle may also provide the answer to HIV infection in humans. Bovine Leukemia Virus (BLV) was first observed in cows in 1969 when researchers were stymied by a mysterious virus in the blood of sick animals. The unusual virus was something new—a retrovirus, one that exists in the body for many years before causing infection. Caused by one rare virus-infected cell, BLV takes a long time to develop, so symptoms are not readily visible. Like HIV, it is a slow virus, and research on BLV has allowed researchers to better understand why HIV antibodies in humans do not prevent illness. Perhaps future work with bovine viruses will unlock the mystery.[28]

Cattle may provide still other health benefits, but vaccination remains the mainstay of human health programs. Vaccination shows how dependent we have always been on the health of cattle. The term was created by Jenner to pay homage to the source: *vacca* is Latin for cow. In respect for Jenner's discovery, Pasteur proposed in 1881 that all protective immunizations against disease be termed *vaccination*, and the idea was accepted. Today smallpox is studied for its potential as a biological warfare weapon, but it is unlikely to be a major threat. Why? Because we still have cattle, and they remain a reliable source of cowpox lymph. If inoculated with cowpox, a calf could supply reliable lymph to vaccinate hundreds of people, in the same way Jenner did. And if cowpox were unavailable, variolation with human smallpox lymph would still be possible; historically the survival rate from variolation ran about 95 percent. But cattle, as our resource against the biological use of smallpox toxins, must be protected. As a species they cannot be bred too narrowly, and their own health must be maintained naturally, without contamination. As a pool of potential vaccine for humans, cattle must be protected from threats such as Mad Cow disease and *e. coli* infection.

Healthy cattle remain our bulwark against disease, but only if we treat them responsibly. Edward Jenner, who owed his success to a stable cow, never forgot that. In 1813 Jenner sent hair from the tail of the famous cow that had supplied his original cowpox matter to another physician, Dr. Richard Worthington. Worthington had been a supporter early on and had encouraged Jenner to self-publish his *Inquiry* when the Royal Society had turned it down. Jenner wrote Worthington that he had "kept the Cow till she died from age. . . . The Hair grew on the Tail of the Cow that infected the Dairy Girl, Sarah Nelmes from whose hand the Matter was taken that spread Vaccination thro' the world. . . . The Cow was a Gloster with a dash of the northern, and a famous milker."[29]

One might call her the cow that proved Malthus wrong.

13

Mad at Cows

Farmers have been accused of encouraging cannibalism in cattle, of being cheap and ignoring the basic moral precepts because they were greedy. But most of them were unaware of what the feed was, because there was no indication of what made up the protein constituent.

—Robert Forster, National Beef Association, London, October 2000[1]

IT IS AN IDYLLIC SETTING in central England, where villages still number only a few thousand people and thatch-roofed cottages and patchwork fields stretch into the distance. Gaggles of geese dabble in brooks, and narrow country lanes cut their way through swaths of green grass. In the eighteenth century, Robert Bakewell's farm, Dishley, made Leicestershire the center of attention for his revolutionary animal breeding techniques. More recently the village of Queniborough, six miles north of Leicester and within twenty miles or so of Bakewell's long-ago farm, has been the focus of world attention. The village was linked to the first "cluster" of deaths from what might become an epidemic with the potential to wipe out a whole generation. Mad Cow disease and its human equivalent, Creutzfeldt-Jakob disease, first made their appearance in Queniborough.[2]

In July 2000 an urgent government investigation began when five cases of a new form of Creutzfeldt-Jakob disease were linked to the village. Local residents were bewildered and frustrated that the town was the focus of the era's most horrifying illness. One resident quipped to the battery of reporters, "What are they all doing here? Are they waiting for people to fall over on the ground and start twitching?"[3]

Two of the victims lived in the village: one worked on a farm, the other had regularly eaten meat purchased at the town butcher shop. David Clarke, the proprietor, ran an award-winning shop—situated halfway down a narrow main street that links the village post office with a beautiful thirteenth-century church at the other end of town. Reporters swarmed the tiny streets and found the local residents "confused . . . greeting questions with defiance and uncertainty." Authorities began looking at the period between 1980 and 1985, when the victims may have contracted the disease. "We have a group of patients, a cluster related in time and space," Philip Monk, the Leicestershire Health Authority consultant explained, "and it is unlikely that this has occurred by chance. Something has happened very locally and we need to find out what it is."[4]

Three of the deaths were linked to Queniborough; two others were nearby. Eighteen-year-old Stacey Robinson died, leaving behind a young child. She had moved away from the village but had been a former resident. Glen Day, a thirty-four-year-old resident of Queniborough, was a farm worker when he fell ill with the disease, and the third victim, Pamela Beyless, twenty-four, did not live or work in the village but often visited her family.

Others had succumbed in the most mysterious illness of modern times—seventy-five deaths in Britain alone by mid-2000. It was a new variation of the elusive CJD, thus called nvCJD. Symptoms are hard to detect. At first the signs are depression, anxiety, and sensory disturbances (vision or hearing affected); the neurological changes are evident when the person has trouble keeping his balance. As the disease progresses, which can take from a few weeks to eighteen months, weight loss, insomnia, and irritability take hold until the person can no longer function, slips into a coma, and dies.

CJD is far more difficult to understand than HIV-AIDS. Unlike viruses or bacteria, CJD is caused by prions, normal proteins, found throughout the body tissues of humans and animals. For some reason, normal prions sometimes transform into abnormal ones that cannot be killed with high temperatures, chemicals, or even radiation. The prions change the surface of the brain's tissue, riddling it with tiny holes. About one person in a million contracts "sporadic" or naturally occurring CJD each year. It had been around most of the twentieth century: one of those obscure, little-known diseases that affects few people, most of them elderly, so it was not at the forefront of national concern. About three hundred Americans have been dying from it every year, most of them past the age of fifty. When a variation of the disease began appearing in young people in Great Britain, however, attention turned to possible infectious agents. Cattle infected with a bovine form of the disease had folded prions in their brains too, and it is believed that when people ate their meat, the disease was passed on to humans.[5]

Since nvCJD was discovered in 1996 it has been a complete mystery; scientists cannot predict whether a future epidemic of gargantuan proportions is building or if the threat is declining. The incubation period is unknown, making it difficult to predict the future. Are we at the beginning of an epidemic that may wipe out an entire generation? "If the incubation period is ten years, then we are in the middle of an epidemic," Dr. Graham Medley, a CJD expert from Warwick University, says. "If it is thirty years, then we are only at the beginning. We have never seen the disease in humans before, so we do not know at the moment."[6] It could be a death threat for hundreds, or hundreds of thousands—no one knows.

Victims of nvCJD were neither infants nor elderly. Of those who had died by March 2000, eight were between ten and nineteen, twenty-three were between twenty and twenty-nine; fourteen were in their thirties; three in their forties; and four in their fifties. Clearly there was an age-related risk factor. Sir Richard Southwood, chair of a government inquiry in 1989 to assess the risks to humans from the epidemic of BSE among British cattle, believed that vaccines containing some sort of infected material were a greater risk than eating beef. But his warning was never made public; the Committee on Safety of Medicines, similar to the U.S. Food and Drug Administration, felt it had a duty not to cause a public panic.[7] So the public was led to believe they could contract the disease only by eating infected beef.

Robert Will, director of the British government's CJD Surveillance Unit, has suggested a new explanation for the number of youthful victims of CJD: he points to food processing as a potential culprit. In the 1980s, when the disease cases were believed to be contracted, meat was being mechanically extracted. Carcasses were sprayed with high-powered water jets to remove every shred of tissue, which was then ground up and used in cheap meat products. School lunches and baby foods were most likely to include the cheap meat product, but it was also included in burgers, sausages, and processed foods. "One possible explanation for the age distribution," Dr. Will explained, "is that young people tend to eat these products more than the adult population."[8]

But not all the British cases can be blamed on eating beef. Iatrogenic transmission—infection from medical procedures—has been responsible for many cases. Medical procedures using infected brain tissue from other humans have passed the disease around the world. At least twenty-three CJD deaths have been caused by infected brain grafts. Six people in Britain and several in the United States resulted from receiving tissue implants taken from infected corpses. Victims in Britain died between four and eight years after tissue transplants that inadvertently used infected tissue. Sixty-five people in Japan have been infected by the same technique. Corpse tissue, commonly used in brain surgery, has been used in many countries for several years. When the connection was made in the early 1990s, Britain banned cadaver brain tissue in surgery. Japanese officials waited longer, not withdrawing cadaver tissue from medical practice until 1997. The Japanese cases are all linked to tissue imported from a German firm; five of the British victims were linked to German medical products too. Dr. Will suggested that "vats containing brain tissue from various corpses had come into contact with some from a single infected cadaver." The practice could have—should have—been altered once CJD made the news. "Everyone says that they did not know and had no idea," said Tetsuji Abe, the lawyer for a Japanese family suing the German manufacturer, the Japanese distributor, and the Japanese Health and Welfare Ministry. "But if they have learnt anything from AIDS, they should have known."[9]

HORMONES, produced in the glands, are responsible for growth, sexual maturation, reproduction, even digestion. Early experiments in the 1920s used the insulin from cow pancreases as a drug for diabetics. That success led scientists to look at other endocrine glands, such as the pituitary gland, a tiny, bean-sized organ that sits behind the bridge of the nose. The pituitary gland was discovered to be responsible for growth, and like insulin from cattle, a bovine source was first explored. But bovine pituitary growth hormone had no effect on people, so researchers looked at human pituitary glands, finding that extracts taken from cadavers worked on children.

In 1958 an American scientist discovered the hormone responsible for human growth, and medication was quickly made available to children who would have become dwarfs without it. The National Institutes of Health supplied the drug free to American pediatricians. For twenty-two years it was administered to more than 8,000 children who had genetic dwarfism. The children grew, everyone was amazed at science, and decades passed. Then, strangely, the adults who had received the medication in childhood began showing odd symptoms: they drooled, they staggered, their personalities changed. After only a few months of these symptoms they sank into comas and died. Upon autopsy their brains were found to be riddled with holes, like a sponge. By the spring of

2000, more than 125 victims who had received the growth hormone as children were dead. The toll continues to climb; there is no way to cure or halt the effects of the medication.

The victims were infected by contaminated hormone, taken from cadavers that were unknowingly infected with CJD. It does occur naturally, in about one in one million persons. "It was an experimental treatment," Jane Demouy, of the National Institutes of Health said. "People signed informed consents."[10]

But the NIH knew there was a danger of contamination in 1978, seven years before the first CJD-related deaths from growth hormone appeared. Scientists supplying the medication under contract to the NIH chose the cheapest, least labor-intensive manner of processing the hormone, rather than a safer alternative that had been developed in Sweden. Swedish researchers had noticed antibodies developing in American children who had been given the drug, suggesting there was something about the medication that caused an immunologic reaction in the body. They refined Swedish production to screen the drug for purity and eliminate contaminated tissues. Eventually NIH investigators estimated that about 140 infected glands may have been used. Currently human growth hormone is synthetic and safe.[11]

CJD cases have also resulted from corneal transplants and contaminated brain electrodes, and a few cases have tenuous (but unproven) links to dental procedures. A few pathologists have become infected, presumably by cadavers.[12] If it is a random disease, affecting 10 percent of the U.S. population, it would mean that thousands harbor the infection. If they were to become organ/tissue donors upon their deaths, their tissue has potential to spread the disease to others.[13]

Blood products are thought to be another avenue of transmission.[14] Alarmed by how seemingly easy transmission may be, U.S. blood banks have barred donors who lived in Britain during the 1980s and have considered excluding blood donors over the age of fifty due to fears they may be harboring CJD.[15] No one can predict how many people may currently be incubating the disease or its variants, nor can anyone at this point determine whether it can be transmitted through blood transfusions or in blood products. Knowing how easily HIV-AIDS is transmitted through blood, however, the scientific community is concerned. Blood donors who do have CJD are barred from donating, but the disease has such a long incubation period, many who may be infected have no way of knowing it.

Ironically, because of fears that blood transfusions during surgery may transmit HIV, blood substitutes derived from cattle blood have been in development for more than ten years. Biopure, a Massachusetts company, hopes to market a blood substitute in the near future. Their product is made from blood obtained in slaughterhouses and then purified by a patented process.[16]

SCRAPIE, the first recognized spongiform encephalopathy, was recorded for the first time in British sheep in 1732. No one knows how it started or where it came from. The first case of scrapie in a cow occurred in France in 1881. Meat by-products were first used in pig feed in 1865, and by 1900 meat by-products from slaughterhouses were being used in ruminant feeds in Britain. Twenty years later H. G. Creutzfeldt published the first case documenting "Creutzfeldt Disease" in a human. The following year A. M. Jakob reported four cases of the disease, which became known as Creutzfeldt-Jakob disease. During the 1920s meat and fish meal was acceptable feed for dairy cattle in Britain; by 1928 it was used in U.S. cattle feed and by the late 1930s in Australian dairies.

The widespread use of meat and bonemeal in cattle feed is now believed to have infected the cattle who ate it. High-protein livestock and pet food is made from rendered animal remains, heated at high temperatures, then dried to create a powder. Renderers process animal remains from slaughterhouses, kennels, dog pounds, and highway departments that must discard road kill—any dead animal that must be disposed of. It is believed that processing sheep infected with scrapie, and employing a less expensive type of processing which used lower heat temperatures, allowed contaminants into the meal, which then infected the animals who ate it.

Because the new-variant CJD cases in humans in Britain appeared a few years after the bovine spongiform encephalopathy epidemic hit cattle, eating meat from BSE-infected cattle was suspected to be the cause. Once the source was determined, legislation restricted such feeds, and BSE among cattle declined. In the United States, risk of BSE transmission is low, according to the American Medical Association, because it has not been shown to exist in this country, and regulations prevent the entry of foreign sources of BSE. Not everyone agrees, and many critics point to the prevalence of Alzheimer's disease as a probable mutation of the disease appearing in the U.S. population.[17]

In spite of its continued link to cattle by the media and government, bovine spongiform encephalopathy has been around at least since the early twentieth century. By 2000 it had made its way into both domestic and wild animals as well as human populations. "Mad Cow" would remain the popular label though, in spite of a growing number of scrapie-infected sheep and "wasting disease" among North American deer, antelope, and elk. All are forms of spongiform encephalopathy. Mink too have been infected with the disease, called TME, or transmissible mink encephalopathy. Because mink were fed remains from "downer" cattle which were slaughtered and rendered into animal feed, a long-undetected level of BSE may already be present in U.S. mink herds. If downer cows—those too sick to stand up or even rise from the ground—are sick because they have BSE, it would explain why certain mink farms where such feed was used had outbreaks of TME among ranch-raised mink. TME is similar to scrapie in sheep and BSE in cattle. Five outbreaks of it have occurred in the United States, beginning in 1947, involving eleven mink farms. Mink are not ruminants like deer and cattle, which have different digestive systems, but they are commonly fed diets of rendered animal remains.[18]

What makes CJD so extremely dangerous is that it cannot be washed, sterilized, or sanitized away. CJD prions on an inanimate object were tested and found to remain infectious three years later. Fear of contamination of persons handling infectious brain and nerve tissue is valid. In 1999 researchers discovered that the disease is present in tonsils and appendix tissues, making surgical procedures likely to pass infection to operating room and hospital employees who handle the tissues.[19]

While cattle, and concomitantly beef and dairy products, have been targeted as a source of CJD in humans, accumulating evidence points to something else as a causative agent. If something in the food is poisoning people, there would be many more incidents; and if it had a genetic component, there would be cases within families or among relatives. Neither is happening. One interesting article in the British medical journal *The Lancet* discussed an epidemic of transmissable spongiform encephalopathy in sheep and goats in Italy. In twenty outbreaks there between August 1996 and October 1997, there were 390 infected animals. Eight flocks were affected, and it was determined that the disease did not come from their feed; eight flocks had never been fed commercial feedstuff. Goats do not naturally contract the disease, so it must have been an accidental infection of some sort. Researchers noted that in 1995–1996 the same animals had been vaccinated against contagious agalactia, with a drug made of brain and mammary gland tissues of sheep infected with Mycoplasma agalactia. The only other outbreak of transmissible spongiform encephalopathy in small ruminants (goats, for example) occurred in Scotland in 1935, after a contaminated vaccine was used against louping-ill (a tick-borne fever). There is no way to know if that incident set off human infection with CJD as records were not kept at that time.[20]

One tantalizing hypothesis has emerged in Britain, where Dr. Anne Maddocks, at St. Mary's Hospital, the largest teaching hospital in London, contends that animal feed was not the culprit behind the Mad Cow disease of the 1980s. She points to the common factor in both human and bovine cases: the use of pituitary gland extracts as growth-promoting hormone agents. Children had been infected by medical use of cadaver pituitaries that were infected with CJD, and so were cattle. Maddocks points to the wide use of pituitary extracts taken from cattle that were used as growth agents by farmers for their cows. Dr. Albert Parlow, research professor at the UCLA School of Medicine, admits that Maddocks may have a good point: "The extracts of pituitary glands that were used in these animals were very crude—far more crude than the human growth hormone preparations."[21] Certainly the widespread use of animal growth hormones taken from the pituitary gland, which sits at the front of the brain, could explain the incidence of more than a million cases of BSE in cattle in Britain as well as the sixty-seven cases of new-variant CJD that appeared in young people who had become infected by tainted cattle products. The human cases and the cattle cases are "mirror images," according to Dr. Maddocks, and both were iatrogenic, caused by attempts to stimulate growth using infected tissues. Bovine hormones were used for cosmetic reasons as well, and resulted in the death from CJD of a French bodybuilder and possibly another victim in Washington state who used hormone supplements for muscle-building.

"The theory is simple," Dr. Maddocks told a London *Observer* reporter. "The promiscuous use of pituitary hormones in cattle led to BSE in the same way that they led to CJD in humans. The timing of the deaths in cattle and humans who were exposed to pituitary hormones is very compelling." Indeed it is; the first case of new-variant CJD appeared before the first case of BSE in Britain. Joanna Wheatley, a former researcher, claimed that abbatoirs (slaughterhouses) were selling pituitary glands to veterinarians and researchers. Cows could have contracted the disease through contaminated brain extract in their hormone injections, and these infected cattle were later "recycled" back into the national herd when carcasses were used as feed or in subsequent bovine medicine. The cycle, once started, was difficult to stop, as one generation of cattle infected the next.[22] Even animals in British zoos were eventually included in the epizootic; a nyala and a gemsbok, both African antelopes, and a lion died after contracting spongiform encephalopathy from feed containing infected animal by-products.

Other theories and hypotheses abound, and fears of contamination seem omnipresent. One experiment with larvae from common houseflies found that when flies were fed with the brains of scrapie-infected hamsters, the larvae from those flies later infected hamsters who ate *them.*[23] One veterinary scientist has proposed a new theory, that the infection comes from a wild animal, likely a rodent, that was rendered into cattle feed in England.[24] At the extreme, perhaps, two British scientists point out that diseases of plants and animals may escape from comets as organic particles that fall to earth. "Small particles of bacterial and viral sizes descend through the Earth's stratosphere mostly during the winter months," they write, "and we believe that the nearly unique English and Welsh practice of out-wintering cattle explains why BSE hit English and Welsh farms more severely than elsewhere." Once infected, a few cattle entered the food chain when they were processed into protein meal and fed to other cattle.[25]

We cannot know how or when the Mad Cow disease epidemic will be fully understood, but the scientific world will no doubt discover major breakthroughs, just as they did with the cattle plague in the nineteenth century. The epizootic of cattle plague, more accurately rinderpest, hit British cattle herds in 1864 and 1865 with a fierceness not known since. The disease was the main public and political issue of its day. More important, its consequences laid the groundwork for understanding the germ theory of disease, thus its impact on medical science was profound.

Rinderpest originated in the steppes of Russian Asia and had threatened Europe since the eighteenth century. It arrived in Britain in June 1865, spread through the dairies, and reached most parts of the country in just six months. Like the twentieth-century experience with Mad Cow disease, the government was slow to act, which allowed many more animals to become infected.

The cattle plague was met with disbelief by the public, who rejected the idea of contagious disease. Everyone, including newspaper editorial writers and well-educated municipal leaders, believed it was "spontaneously developed," that "some thoroughly well-educated medical men" should investigate and find a cure. The Cattle Plague Commission found that its orders for cattle isolation and slaughter met with opposition from the medical profession, who believed a cure could be found. A bevy of folk remedies were tried: homeopathy, hydropathy, bleeding, even alcohol and opiates. Inoculation was tried too, but it succeeded only in spreading the disease.[26]

According to John Fisher, "Virtually all the great names in British medicine of the 1860s acquired clinical or at least firsthand knowledge of rinderpest," yet they could not accept the idea that if an animal (or person) were isolated, a disease would disappear on its own.[27] The idea of disease causation was not very interesting; physicians were more intent on finding cures. The belief that miasmas were the source of disease was a long-held, seldom challenged view. Miasmas were thought to rise up from swamps or putrid matter. Since all disease was thought to originate from miasmas, the only recourse was to clean up the environment or to discover cures in the form of elixirs, nostrums, and such. After all, public health successes were based on cleaning up filth, not quarantine.

The cattle plague revealed that "climate, altitude, dirt, or miasmas" were not as significant as previously thought. And studying how cattle experienced an epidemic of contagious disease was much easier and clearer than watching the same effect on the human population.[28] Because rinderpest did not infect humans, it could be examined at leisure and without emotion. The cattle plague caused the British scientific community to look hard at its accepted theory of disease and ultimately eased the recognition of germ theory.[29] Researchers first looked at chemistry, but eventually, with the microscope, discovered biological causes. Minute particles, referred to as "germ" or "growth," because "no better expressions can be found," were believed to be specific to the disease. The germs multiplied quickly and could be dispersed in the air. In the future, "germs" were more commonly accepted, and the idea of miasmas faded away.

Estimates of more than 400,000 dead cattle, and the speed which the disease hit, made it important news in the *Times* in 1865. Because financial losses were so great, the government was quick to provide money for research and legislation to limit the spread of the disease by quarantine, forced killing of infected cattle, and reduced imports and exports. Interestingly, a cholera epidemic during the same years caused 15,000 human deaths, but it did not receive nearly the attention that rinderpest did. Cholera hit the poorer classes hardest while the cattle plague affected landowners and the upper classes.

Today's Mad Cow disease will no doubt again shift medical and scientific thinking in a new direction. Already the concept of infectious misfolded proteins that are believed to spread through the brain tissue is entirely new. Whatever conclusions are finally reached regarding BSE and CJD, we will have a new paradigm of disease that we had not been prepared for.

One problem that will have to be met is how to dispose of the millions of tons of offal and animal carcasses that have so handily disappeared into the renderer's vat and emerged as "protein meal." In Britain one power plant is already burning the carcasses of animal remains to generate electricity. They had been using poultry waste but switched when so many BSE-suspect cattle were slaughtered. The plant burns 250 tons of cattle carcasses and bonemeal a day, generating enough electricity to power a small town.[30] In France, banned protein meal animal feed has been stockpiled and is being burned as fuel at cement factories.[31]

The rendering industry is experimenting with using animal byproducts as dust-control applications for unpaved roads and as a component of "biodiesel" fuel. Their biggest market, however, is still animal feeding, and commercial fish farms may be the last market for rendered animal protein meal. As Dr. Dominique Bureau, from the University of Guelph, Canada, remarked at the annual meeting of the Fats and Proteins Research Foundation: "The belief is fish need to eat fish. But fish don't care."[32]

But there is also evidence from the aquaculture industry that unprocessed or improperly processed fish protein may be problematic as a feed for fish. One ailment of trout and salmon whose introduction into the United States may have been linked to this problem is whirling disease, named for the uncontrolled behavior it provokes. Evidence is circumstantial, but parasite spores may have been introduced in shipments of frozen fish from Denmark during the 1950s and 1960s that were later used as fish food. Although microbiologist and fish disease specialist Jerri Bartholomew of Oregon State University points out that there are other, more likely routes of dissemination, the resistance of the parasite spore to freezing and its location in fish cartilage does make this hypothesis plausible. Bartholomew notes that "there is ample evidence in the fish feeding industry that demonstrates the problems that can occur when insufficiently processed fish are fed in hatcheries."[33]

IF MAD COWS did not present enough of a problem for British farmers and beefeaters, foot-and-mouth disease appeared in 2001. It had not been seen in Britain for decades when the first case was identified on February 21, at an abbatoir in Essex. The next day a sick bull was identified in southern England, and by February 24 affected animals were being slaughtered at eight sites across the country. Veterinarians had identified cases from the northeast to the deep southwest, in a slash across the English countryside. Highly contagious, the airborne virus can be passed on one's infected shoes. Within days, British farming areas were in a virtual lockdown, with no travel allowed.

British farmers had experienced zoonoses before and recognized that the only solution for highly contagious foot-and-mouth disease was government enforcement of quarantine and slaughter. The British public grew shocked and dismayed at the macabre piles of livestock being burned. How could it happen in a country so alert to cattle health after the BSE experience?

Theories abounded, but as of this writing none had been confirmed. Pig feed, made from infected rendered animals, may have been imported from Asia where foot-and-mouth disease is still endemic. But cattle came down with the disease at the same time as the first pig cases, and rendered animal by-products were no longer being fed to cattle in Britain. Other investigators looked to Argentina, where a few cases of foot-and-mouth disease had occurred in the summer of 2000 along the remote border with Brazil and Paraguay. Lambs imported from Argentina had initiated the last

British epizootic of foot-and-mouth in 1967; perhaps history was repeating itself. But tests revealed that the strain in England and the recent cases in Argentina did not have the same DNA. Wild birds, migrating from Africa each spring, were also suspect, just as they had been in 1967. But there is no evidence they carried the virus from Africa where it is still endemic.

Another proposed scenario that viral contaminants might have escaped from a U.S. research facility on Plum Island, north of New York's Long Island, the only place in the United States where research on foot-and-mouth disease is allowed. Suspicions arose that the virus blew across the Atlantic. Although the virus survives well in cold, moist temperatures, if it indeed traveled that distance it would certainly have affected livestock in Ireland first.

And thoughts kept returning to how effective the virus might be as a biological weapon. Had some terrorist sprayed the English countryside with foot-and-mouth virus from an airplane? The infected areas cut a straight swath, north to south, across the country, not scattered randomly like Mad Cow disease nor centered at coastal shipping centers where imports arrived. Some observers pointed a finger at Iraq, because United Nations Special Commission (UNSCOM) inspectors had battled with the Iraqi government for years over inspection of suspected bioweapons facilities there. One of the most contested sites was the Daura Foot and Mouth Disease Vaccine facility near Baghdad, where UNSCOM worried that anthrax was being harbored. The first incidents of foot-and-mouth in England appeared only five days after a U.S.-U.K. bombing strike on Baghdad on February 16, 2001. Because the disease has a two-to-fourteen-day incubation period, many wondered if it was the result of Iraqi retaliation.

Biological attacks on livestock are nothing new. They were suspected during the American Civil War when Confederate horses were hit with glanders; again during World War I when German agents infected horses and cattle with glanders and anthrax on U.S. docks before they were shipped to the front; and in World War II when the Japanese used bioagents on animals and people in China. According to German historians, the Nazis were prepared to use foot-and-mouth virus against Britain in the final months of the war. In fact it was during a 1938 epizootic of the disease in Germany that officials recognized how devastating it would be to British livestock, which had no immunity to the disease. Testing was done using a spray tank attached to a bomber, successfully infecting a herd of Russian reindeer. But the war ended before the tactic could be employed.

While the cause of the recent British epidemic may never be found, it does create apprehension over a new threat: agroterrorism. Agricultural bioterrorism is recognized to be a growing area of vulnerability; the U.S. Defense Department has cited it as a prime concern. Foot-and-mouth disease, and other plant and animal pathogens, have been recognized for their potential as weapons and are being taken very seriously by the defense and intelligence communities. While foot-and-mouth disease does no harm to humans, it can wreak havoc on a nation's food supply, export market, and economy.

14

Carnivore Culture

> William Stark, a healthy 29-year-old physician in the eighteenth century, tried to figure out how diet affected health. He ate carefully weighed portions of bread and water with the addition of other foods one at a time. Within a few months of beginning the experiment, he sickened and died from what today would be recognized as severe vitamin deficiencies.
>
> —Wayne Rasmussen, *Agriculture in the United States*[1]

INEVITABLY any discussion of our connection to cattle must address the fact that they taste good. Beef has been a dietary staple for centuries, and for good reason. It is delicious, easy to prepare, and makes one strong and healthy. But today eating beef (or choosing not to eat it) has become a bit problematic. Is eating beef safe? Is it good for one's health? Is it too greedy? Is it cruel? Nothing riles a group as quickly as discussing religion, politics, or whether eating meat is a good idea.

It is not surprising that early humans ate considerable amounts of meat, in fact recent archaeological research reveals that their diet was almost entirely meat. And they were strong, tall, and fairly disease free. Today there is a growing movement to eat like Paleolithic people did. A popular high-protein diet plan is based on anthropological and archaeological research on the eating habits of early humans. The Stone Age diet, which humans ate for 2.6 million years (plant agriculture was adopted only 10,000 years ago), now appears to be optimal for human health. Early people did not eat cereal, they ate large amounts of meat, particularly the fat, along with berries, nuts, and roots. They ate little if any starch and did not consume sugar in the sucrose form we do today. Focusing on the eating habits of cavemen is something new; for most of human history we have believed we are advancing toward a sort of Darwinian apex in nutritive knowledge and behavior. Not so, physiologists and physicians now tell us. Grain-based diets are being blamed for blood insulin imbalances, reduced mineral absorption, and many chronic diseases.

The hunter-gatherer diet—the "caveman diet"—has a growing number of proponents. Nibbling on sardines, beef jerky, and grapes while one answers e-mail messages is becoming quite popular. Several "paleo diet" books have appeared in the wake of the groundbreaking 1988 book *The Paleolithic Prescription*, written by a team of doctors and anthropologists, Boyd Eaton, Marjorie Shostak, and Melvin Konner. When they first published a technical paper on the subject in the *New England Journal of Medicine*, the authors found themselves the butt of jokes and cartoons. Nevertheless the medical community paid attention and even welcomed the new perspective. Researchers have begun assessing whether the chronic and deadly "diseases of civilization"—heart disease, strokes, cancer, diabetes, hypertension, cirrhosis, and similar disorders—are related to eating a diet incompatible with our genetic and biological adjustment. According to nutrition author Jo Robinson, beef contains vital Omega-3 fatty acids in large amounts *if* the animal was fed grass or hay. Feedlot beef, fattened on grain and other feeds, has fewer nutritive fatty acids. Pastured beef is more like the free-range game that our ancestors ate.[2]

After the appearance of agriculture, dietary proteins diminished because meat was no longer the main food source. Grains dominated the diet, providing far less optimal protein sources. This inadequate intake of protein "dwarfed" ensuing generations of farming peoples. "This may come as a surprise to people who have, for the sake of health, cut back on the one protein source most readily consumed by hunters and gatherers: red meat," the authors of *The Paleolithic Prescription* note. But they point out that red meat has changed over the ages. Some of the animals that hunters ate no longer exist, such as mammoths and giant sloths. Others do, but we simply do not eat them, such as horses. The animals that we continue to dine on include cattle, of course, along with sheep, goats, and pigs. Paleolithic peoples consumed at least 35 percent of their calories in protein, two to three times what USDA nutritionists recommend today. Boyd Eaton calls modern America's grain-heavy, low-meat diet "affluent malnutrition," and recommends that 20 percent of our daily caloric intake come from animal protein sources.[3] Eating lots of meat and fish, some fruit and root vegetables, and few or no grains or dairy products—the caveman diet—resembles the USDA food pyramid turned upside-down. It contradicts everything nutrition experts have been telling us since the 1960s.

Since 1992 the USDA's food pyramid has set the standard for determining what people should eat. (Before that, the rubric was a model of four food groups.) The pyramid suggests six to eleven servings of bread, cereal, rice, and pasta per day, and two to three servings of meat, poultry, fish, dry beans, eggs, or nuts. Dr. Michael Eades, who has written several high-protein diet books with his wife Dr. Mary Dan Eades, points out that this pyramid is identical to the proportions of grain to protein that are present in commercial hog feed. Such a heavy reliance on grain-based feeds results in obesity—desirable in hogs but not in humans.

People may be perusing diet books, but research from the National Cancer Institute reveals that Americans rely on fortified breakfast cereals as sources for more than half their essential nutrients. Children consume fortified cereals and fruit drinks that contain little real fruit juice for a large part of their essential vitamins and minerals. Children also eat large amounts of foods with little nutritional value, making them feel full but not providing enough vitamins and minerals. Major sources of protein in U.S. children's diets turn out to be poultry, ready-to-eat cereal, and pasta. They eat little or no red meat.[4]

Recent research into heart disease has overturned ideas that dietary fat causes cancer and heart disease and that consuming cholesterol is bad for one's health. Heart disease is now thought to be related to viral or bacterial infections. Because new studies show that inflammation and antibodies accompany heart disease and atherosclerosis, research is focusing on how infection plays a role in heart attacks and heart failure. According to research at the University of Washington, white blood cell count seems to be more related to heart disease than blood cholesterol levels.

Research findings on cancer too have exonerated dietary fats. For years women were warned against eating animal fat because it was thought to be connected to breast cancer. A long-term study of nurses and their dietary intake has revealed that consuming animal fats did not cause breast cancer.

Even eggs, long thought to be bad for health because of their cholesterol content, have been reevaluated and found to be excellent sources of nutrition. We now recognize several types of cholesterol, and eggs contain the "good" kind. Red meat too has been vindicated, and physicians are encouraging people to eat more meat—but cut away the fat to save calories.

What exactly makes meat so valuable in the diet? The USDA Nutrient Database lists the following components of beef: protein, calcium, iron, magnesium, phosphorus, potassium, sodium, zinc, copper, manganese, selenium, thiamin, riboflavin, niacin, pantothenic acid, vitamin B-6, folate, vitamin B-12, fatty acids (saturated, monounsaturated, and polyunsaturated), cholesterol, tryptophan, threonine, isoleucine, leucine, lysine, methionine, cystine, phenylalanine, tyrosine, valine, arginine, histidine, alanine, aspartic acid, glutamic acid, glycine, proline, and serine. Most valuable from meat, beyond the protein and iron, are the amino acids, which are vital to functions of the nervous system and brain. Without enough amino acids, brain function deteriorates and, just as in cases of pellagra, dysfunction sets in. Nutritional counselors often urge women to eat more protein because inadequate amounts of it can lead to depression. Perhaps the reason more women suffer from depression than men is due to their eating lesser amounts of red meat.

The favorable effects of animal products upon body development, muscle power, and general health and strength have been compared in two African tribes, the Masai and the Akikuyu. Both peoples have intermarried, so there is some genetic overlap, which eliminates genetic makeup as a variable. The Masai consume a diet high in protein, with plenty of meat and milk, while the Akikuyu eat a vegetarian diet of maize cereals, legumes, sweet potatoes, and greens. The Akikuyu show higher levels of bone deformities, dental caries, anemia, pulmonary disease, and ulcers; the Masai, who eat large amounts of meat and milk, average five inches taller and fifty pounds heavier, and have twice the muscle power while being comparatively free from disease.[5]

TODAY people decide what to eat largely because of advertising, and advertisers tout nutritional benefits, no matter how dubious. Historically, however, people have not chosen their foods based on nutritional value. Without knowledge about the body's metabolism or the nutrient makeup of foods, people have used other beliefs to justify their eating habits. From classical Greece until the sixteenth century, food was classified according to its purported effects on the body. Human beings and their food, and everything else within the cosmos, was thought to be made up of the same four elements: air, fire, water, and earth. Within the body these created the four humours: blood, bile, phlegm, and black bile. People who were sickly suffered from an excess of one or more of the four humours. The way to overcome ill effects was to eat foods that helped balance the internal situation. Foods were categorized as to which of the four humors they affected: hot, cold, dry, or moist. For example, the elderly were believed to suffer from being too cold, therefore they were advised to eat a diet of "hot" foods. Meat was a hot food because it was from animals; so were eggs. Greeks ate plenty of meat; the Romans, who followed them, ate far less. The typical Roman citizen's diet consisted of large amounts of grain in the form of bread, with olive oil, cheeses, and beans.[6]

Europeans raised cattle early on, and their diets reflected it—in what the historian Fernand Braudel has called a "riot of meat." Beef was served boiled or roasted and in the fifteenth and sixteenth centuries was eaten by even the lower classes. Wild game, small animals, and birds made up a large part of the diet when beef was in short supply. In Berlin in 1763, the king ordered a hundred head of deer and twenty hogs brought into the city every week because cattle supplies were low. Braudel describes the "food pyramid" of "carnivorous Europe" in the Middle Ages: "meat in all its forms, boiled or roasted; mixed with vegetables and even with fish, was served 'in a pyramid,' on immense dishes." Meat abounded in the markets and eating houses, and servants expected two servings of meat at a meal.[7]

The cattle trade in Europe at this time was on a grand scale too. Herds of 16,000 to 20,000 animals poured into the cattle fair in Germany, near Weimar. Cattle trade overland and by sea to the slaughterhouses of central Europe amounted to 400,000 head per year. In Paris alone, 70,000 cattle were slaughtered annually. In the sixteenth century, half-wild cattle still roamed Hungary and the Balkans and were rounded up and herded to market.[8]

But beef consumption did not last, and by the seventeenth century, with the rise of agriculture, the poor were existing on bread and gruel. Famines weakened the population, and disease was rampant. The nineteenth-century Irish famine hit that country so hard because by that time the people were existing on little more than potatoes. Weak, sickly, and on a scant, starchy diet, the Irish succumbed to epidemics as much as to malnutrition.

VIGOR, a nervous energy that gave one "get up and go," was believed to be the secret to Western societies' prosperity, and vigor was thought to be the result of a meat diet. In the mid-1800s interest in applying scientific knowledge led to considering what sort of diet made people most energetic and able to perform physical labor. Research looked at the kind and amount of particular foods essential for stamina. The feeding of prisoners and paupers in Britain was an example of such research, as the government sought to feed them adequately (both classes were expected to continue to do physical labor). But their fare could not be luxurious, as that might undermine social goals. They had to be fed, as it was un-Christian to let them die from starvation, but they could not be provided the same foods the poor working family ate. Where was the punishment (for prisoners) or the chastisement (for the poor) in that? Thus prisoners were fed bread and gruel (oatmeal boiled with water or skimmed milk) for the first two weeks of confinement, followed by six weeks of meat and potatoes twice a week, and after that four times a week with soup at other times. Hard labor—the core of punishment—necessitated strength, so the prisoners had to be fed well enough that they could perform adequately in order to endure their sentence. They were usually put to work on stair-step treadmills which operated simple machinery, or handed a pick or shovel in rock quarries or coal yards.[9]

Protein, called "animal substance," was first discovered by Justus Liebig, the leading German chemist of the nineteenth century, who believed it provided the sole source of energy for muscle movement. Liebig marketed his own soup-based, food-supplement products, which appealed to those who abhorred or could not afford to eat meat. He sold Liebig's Extract of Meat—bouillon, really—prepared from finely chopped lean meat boiled in water. The water was then strained off and sold as "extract." Leibig pointed out that meat contained creatine, not found in plants. If one chose not to eat meat, a plant-food diet could be supplemented with the "extract of meat" in order to supply the body with creatine and other important substances.[10] Liebig's ads were misleading though, implying that one ounce of extract was equal to the nutritive value of two pounds of meat, and that it would help invalids return to health. His extract was highly popular in the United States, and copycat marketers followed suit.

Florence Nightingale discussed the purported benefits of beef "tea" in her book *Notes on Nursing* in 1860, telling nurses that the belief that beef tea was the most nutritive of all tonics was unproven. "Just try and boil down a lb. of beef into beef tea, evaporate your beef tea, and see what is left of your beef. You will find that there is barely a teaspoonful of solid nourishment to half a pint of water in beef tea; —nevertheless there is a certain reparative quality in it, we do not know what." She was quick to criticize vegetarians who had claimed "an egg is equivalent to a lb. of meat, —whereas it is not at all so." She did admonish nurses not to go overboard with their private patients, feeding them only meat as a curative diet, because without vegetables they might develop scurvy.[11]

Early in the 1800s a dietary philosophy had developed in England and later in the United States that it was wrong to kill animals and eat them. Originated by the ancient Greeks, it had continued into early Europe and had included such well-known devotees as Leonardo da Vinci. The vegetarian ideal connected with the Romantic movement, a reaction against the rationality of the eighteenth-century Age of Enlightenment in France. Simplicity was sought rather than the artificiality and decadence of sophistication. Eating plant foods was a symbolic return to an innocent Garden of Eden, eating meat was equated with an "unnatural" way of life.

Utopian perfectionists, religious groups, and upper-income Americans, particularly females, also adopted vegetarian diets that scorned meat-eating. Nineteenth-century health reformers claimed that high-protein intakes were harmful. By the Victorian era, meat-eating was linked with overexcitement and stimulation, which was debilitating and demoralizing—due to meat protein being a source of animal heat. Vegetarian diets were the safer, saner, and more refined mode of living. Meat-eating was culturally equated with physical labor, and middle-class Victorian intellectuals distanced themselves from the laboring masses by turning away from a meat diet in favor of a more sophisticated manner of eating, embracing the idea of vegetarianism.[12]

An anti-meat health movement led by Sylvester Graham (1794–1851) laid the foundation for today's vegetarian diet stance.

Graham was a Presbyterian minister who was hired in 1830 to lecture at the Philadelphia Temperance Society on the subjects of alcoholism, chastity, prevention of cholera, and diet reform. He advocated eliminating meat. He was a huge success; speaking engagements in New York, Boston, and Philadelphia paid him $300 a night (when nurses earned $3 a week). Protesters showed up to heckle his appearances, mainly butchers whose livelihood was threatened and critics who felt he dwelt too heavily on sexual topics.

Graham's simple solution had appeal. Physicians at the time believed that diet was important in regulating the body. Overheating was dangerous, in the form of overstimulation or eating meat, eggs, or spices, which were all thought to be "heating." Fruits and vegetables were "cooling," and preferred for invalids or pregnant women.[13] Graham advised people to eat as little food as possible to maintain energy, and to exclude all meat as well as alcohol, tea, coffee, pepper, and mustard. His vegetarian diet promised to solve many of the problems of the day: disease, crime, social evils, moral corruption.

Graham, who had experienced a sad, lonely childhood (his father died when he was a toddler, and a few years later his mother was admitted to a mental hospital), gravitated to a true comfort food: home-baked whole wheat bread. To him it symbolized home and hearth as well as health. He advocated avoiding store bread and encouraged mothers to bake their family's bread with whole wheat flour as well as love.[14]

In spite of his healthful program, Graham died at the age of fifty-seven. His shoes were filled by young John Harvey Kellogg, a public speaker who continued vegetarian lectures built on a religious platform. The Seventh Day Adventists, a religious group which began in three small churches in western New York state, by 1855 had relocated to Battle Creek, Michigan, where they had adopted healthful living as a keynote. They prohibited alcohol, tobacco, tea, coffee, pepper, and other condiments, and, in the 1860s, meat. The Adventists (who believed the seventh day was on Saturday rather than Sunday, and that the Advent or second coming of Christ was near) built a health asylum at Battle Creek in 1866 where ill followers could be cured without drugs. Healthy living and religion were thought to be enough; no trained medical staff were necessary. But the institution was a near failure until the twenty-four-year-old Kellogg took it over. He had been raised as an Adventist and had been sent to a New York medical school by the church. But he had much grander ideas and set about turning the sometime hospital into a large medical sanitarium. By 1907, under Kellogg's direction and charisma, the place was buzzing—celebrities like Henry Ford and John D. Rockefeller came for the healthful relaxation.

Like Graham, whom he had followed since his youth, Kellogg equated sexual energy and meat with dissipation. He extolled the dangers of sexual incontinence along with what he called "autointoxication" from eating meat. Of his several books, one, *Colon Hygiene*, was based entirely on the human digestive system and meat. In it Kellogg explained that surgeons had resorted to removing the colon because it was believed to be a useless remnant from a prehistoric state. Sophisticated, highly evolved Darwinian thinkers readily accepted Kellogg's views, believing themselves to be evolving beyond the use of their own colons. After all, it was a dirty, disgusting organ, one that clean-minded Progressives might do well to abandon. Kellogg called the colon "the germ of old age" and blamed it for heart disease and arteriosclerosis. English surgeons, he claimed, had discovered that removing the colon cured tuberculosis, rheumatism, and countless other ailments.[15] "That the average colon, in civilized communities, is in a desperately depraved and dangerous condition, can no longer be doubted," he wrote. "The colon must be either removed or reformed."[16] For people who were reluctant to allow surgeons to remove their colon, Kellogg's dietary regime promised to reform the evil organ without resorting to dangerous abdominal surgery.

Ironically, Kellogg used cattle feeding as an example of how diet could be altered radically while maintaining good health. He noted that when grain prices were high in England, cattle were fed treacle, a cheap molasses syrup, mixed with wood sawdust. "The animals readily fatten on this diet and remain in good health," he wrote.[17] Why wouldn't humans do as well on a high-fiber diet?

Eating animal flesh he viewed with abomination: "As ordinarily eaten, the flesh of animals is always in a state of more or less advanced putrefaction, and hundreds of millions of living bacteria are found in every morsel."[18] Bacteria had recently been discovered, and Kellogg cited the bacterial action that caused meat to rot as a danger in the human gut. Physicians in the 1880s thought this putrefaction of meat must occur in the intestines or colon. This was the cause of "autointoxication." Kellogg believed that toxic bacteria remained in the colon where they had been introduced by infected meat.

The solution? Kellogg recommended three bowel movements a day and a diet low in protein and without meat. At the sanitarium this meant multiple daily enemas or "colonics" and a diet of cereals. The Kelloggs (his brother William developed the technique for cereal flaking) created a cereal mix they named "granola" and in 1895 introduced flaked cereals at Battle Creek.[19] John Kellogg also encouraged eating large portions of "Bulgarian ferment"—yogurt—made from sour milk.

While proponents of healthful Christian living adopted some if not all of Kellogg's beliefs about meat, the idea that meat equated with vigor persisted among many people. Meat-eating, hence national "muscularity," was equated with success. Italians were considered "idle, apathetic, and absolutely degenerate" by a writer who attributed this condition to low protein consumption. The same writer claimed that workmen on a French railway line were less productive until fed the same portions of roast beef that English workers on the line ate. Beef-eating was linked to imperialist success: "A few tens of thousands of well-fed English carnivores were able to hold in subjection a hundred million vegetarian Hindus," explained an economist in 1896.[20]

Liberal amounts of protein were eaten by the leading nations of the world at the beginning of the twentieth century. England,

France, Italy, and Russia as well as the United States all adopted the belief that "a moderately liberal quantity of protein is demanded by communities occupying leading positions in the world." "The negro and the poor white of the South, the laborers in Southern Italy, all partake of diets relatively low in protein. That their sociological condition and commercial enterprise are on a par with their diet, no one doubts. . . . When people accustomed to a low protein diet are fed on a higher protein plane, as is the case when southern Italians come to America, their productive power increases markedly," a writer noted in 1906.[21]

The protein controversy was bolstered by reports from British physicians in India who compared the Sikhs, a muscular, vigorous people, with the Hindus in Bengal. The Hindus were characterized as flabby and lethargic, and poor workers. The difference was due to diet: the Sikhs ate meat and milk (about 135 grams of protein per day) while the Bengali's rice diet lacked animal protein, and included less than half the protein the Sikhs received. This led researchers to question whether the level of protein necessary to exist was an amount that could provide necessary strength and stamina. D. McCay, a British professor of physiology in Calcutta in 1912, pointed out that "It is not without reason that the more energetic races of the world have been meat-eaters."[22]

Virility, energy, power—what nation wanted less?

IN 1879 Wilbur Atwater, a Yale professor and founder of the science of nutrition in the United States, concluded the first study of foods and their nutrients. He created an index of the economic values of foods based on a German method of measure. Beans, grains, and potatoes, he reported, were more economical sources of protein than animal or fish. Writing in the *Century Magazine*, which had a prosperous, middle-class readership, Atwater criticized the American poor for spending their money on meat when less expensive protein sources were available. If the poor spent most of their available funds on food, they had very little left for housing or self-improvement.[23] This poor management of resources was used to explain the lack of upward mobility among the poor and the working class. Working-class Americans were advised to spend their money more wisely, passing up cuts of meat in favor of cheaper potatoes.

Atwater's work reinforced the idea that the poor were to blame for their sorry state—if only they would economize a bit and not expect to eat like their betters, they would eventually rise out of poverty. The hidden agenda behind studying nutrition as a science during the Gilded Age may have been to figure out how to persuade workers to live on less money.[24]

Atwater set the protein standard for physically active men at 125 grams per day. Working people were urged to buy cheaper cuts of meat that were just as nutritious as more expensive ones, so that they could meet the requirements.[25] But that was not so easily done on the wages of the era. Russell Chittenden, an established physiological chemist, felt that Atwater's standard was too high, and he studied the physical performance of volunteers who consumed only about half the recommended standard of protein for six months. His subjects reported no ill effects; but others were not persuaded that such a diet could be safely maintained for longer periods, because it seemed a general rule of life that those who could afford to do so ate more protein, and that the poor who did not were generally lacking in energy and initiative. Chittenden tried unsuccessfully to argue that the deficiencies of "poor" diets were more likely to be those of unidentified trace nutrients than those of protein, and that it was affluence that *caused* people to eat a "rich" diet, not vice versa as his critics claimed.[26] What made Chittenden's low-protein diet more difficult to accept was that experiments in two German laboratories in the 1890s that had fed dogs low-protein diets had resulted in the dogs remaining healthy for weeks, after which they lost their appetite and inexplicably died. Experiments in Finland showed similar results.

Another 1890s theory proposed to lessen the amount of necessary food. Wealthy Horace Fletcher promoted his idea that chewing food more thoroughly reduced the amount of food one needed to consume. "Fletcherizing"—chewing food as long as possible—was briefly popular at the Battle Creek sanitarium.

Changes in the American diet were viewed with skepticism by the federal government: they might physically weaken the nation, because diet was related to national security. If soldiers were necessary for defense, fighting men had to come from the ranks of the working classes, and they needed to be adequately nourished. In an era when manpower was vital to defense, both rich and poor had a stake in the nation's diet. By the latter twentieth century, with the use of technology and science to win wars, a nation's protein intake became less important.

The quantity and source of dietary protein was a subject of controversy in Sylvester Graham's era, again after World War I, and in the late 1960s, when vegetarianism returned to popularity. Diet books extolled the virtues of eating plants as a sound, environmentally friendly way to survive. Francis Moore Lappé's book, *Diet for a Small Planet*, published with the Friends of the Earth in 1971, set the standards for eating low on the food chain. The book recommended eating less protein and appealed to the educated middle class who found eating natural products a way to distinguish themselves from "most people" who ate "standard fare." Elites and intellectuals adopted "exotic" foods from peasant diets: bulgur, yogurt, tofu. The creativity involved in putting together meals made of incomplete proteins provided a scientific challenge for college-educated Americans, distancing them from those content to fry a burger—and less capable of understanding incomplete versus complete proteins. Eating unusual foods gave consumers a sense of power in the marketplace as they sought obscure or exotic foods. Beef became known as "red meat," a symbol of waste and excess as well as the fare of a generation past. Roast beef and gravy symbolized a blue-collar lifestyle; quiche or tofu suggested that one was intelligent, thoughtful, and worldly-wise. Macrobiotic vegetarian diets became fads in many university towns, but not everyone knew enough about nutrition to eat safely. In Berkeley, California, incidents of kwashiorkor, a deficiency due to lack of protein, rose.[27]

In an era when concerns about pesticide levels in foods were a real threat, Lappé's message was timely and valuable. But the idea that lesser-quality proteins could be just as healthy allowed food processors to create and fill a market for "natural" products that were essentially junk foods. School lunch menus reduced the amounts of meat in favor of starchier alternatives. Beef was passed over as unhealthy or greedy when the world was getting by just fine on a handful of grains. What scientists had been trying to do for decades—justify feeding the poor a meatless diet—was adopted as a cultural symbol by a youthful generation of middle-class baby boomers.

Theoretically, if Third World peoples could eat cheaper, more available protein sources, they could free up resources that are used in raising livestock. The World Health Organization, with funding from the World Bank, began studying alternative sources of protein in the 1950s. Fish, soybean, peanut, sesame, cottonseed, and coconut sources were studied; added later were genetically improved plants, yeasts and bacteria as "single-cell protein sources," and synthetic amino acids. Among the products developed as a result of these studies was a fish flour called "fish protein concentrate." The idea was that poor children might thrive on the fish meal, a particularly popular idea with governments in Scandinavia, Peru, South Africa, and Chile because of their ready supply of fish. President Kennedy's administration eagerly established fish protein concentrate plants in the United States, but the kind of fish used and the high cost of fuel to process it resulted in the project's demise. The fish used were herrings and anchovies, and the entire fish, offal included, was processed into meal. But the fish fat content was so high that the meal stank and was costly to process due to the amount of energy needed for steaming, heating, and drying. The processing also removed B vitamins. The FDA was unwilling to allow fish protein concentrate as a human food unless it was disease bacteria–free, something that would have added still more to the cost of processing. And the FDA would not allow the entire fish to be processed into meal, yet the small size of the fish made it impossible to clean out the innards. Fish bones contained high levels of fluorine too, considered a health hazard. Protests against fish protein concentrate arose from the dairy industry, which believed it would destroy the third world market for dried milk powder, which had to meet tougher processing standards than fish flour.

A range of other protein sources were examined. Oilseed flours were made by pressing vegetable seeds for their oil content, then drying the bulk. The oilseed flour was stirred in an alkaline solution, then treated with acid to remove the protein. Flours made from peanuts, soy, and cottonseed were also tried but abandoned due to problems with fungus and contamination, high costs, and distribution problems. Single-cell proteins—yeasts, fungi, and bacteria—were tried too. After all, a protein source that grew itself would be an incredible boon.

Tests in animal feeds seemed to work on rats and pigs, but humans had a problem tolerating high levels of nucleic acids which became toxic when large amounts of micro-organisms were consumed. A second problem was that 15 percent of the people in the feeding tests reacted with severe allergic reactions (eczema, vomiting, and diarrhea). A third problem was the potential for production facilities to become contaminated by unwanted micro-organisms, creating high costs for sterilization or vacuum conditions. By 1977 the project had come to a halt. No new cheap protein foods had been created, and more important, the emergency of third world nutrition had faded.[28] Still, U.S. food processors adopted many of the cheap proteins as additives, and today protein isolates are commonly found in processed foods.

In 1975 the United Nations World Food Conference concluded that instead of continuing the search for cheap protein substitutes, widely available foods should simply be fortified with amino acids, protein concentrates, vitamins, and minerals. By 1977 the WHO program that had begun in the 1950s was dissolved.[29] The world's impoverished children were viewed as suffering from a food *quantity* gap rather than a food *quality* gap. The solution was thought to be more calories, not more protein. The real problem—poverty—was not addressed. Many of the countries did have adequate protein supplies, but they were not distributed evenly across the society for political, religious, or economic reasons.

It seemed there was little that American agricultural science could do about the distribution of protein. Food scientists had studied the problem for more than a decade before they found there was no use for their product.[30] The new technologies required so much fuel energy to produce each food calorie that it was unavailable to poor countries. By the early 1980s the hope that technology could create a greater food supply had disappeared. It might have seemed that people could survive on chicken feed, literally, but getting them to eat it had not been possible.[31]

At the same time affluent diets were found to be just as devastating to health as the diets of the poor. Americans were overeating, growing obese as well as sickly. So American concerns shifted from attaining more protein and food energy to eating less. Affluent diets in recent years have focused on *avoiding* rather than consuming nutrients, particularly fat, cholesterol, and calories. This has led to lower consumption of animal protein foods, even to the point where meat is used as a flavoring agent or garnish rather than a staple.

EATING MEAT is now more complicated than ever before, because consumers have many more things to worry about. *The Jungle* taught people to examine how animals were slaughtered and meat adulterated. But livestock farmers had been ignored by consumers, who suddenly began to ask questions. They found that the traditional grass-fed animal had been replaced by scientifically fed animals in large feedlots. Private nonprofit groups began to question the use of animal feeds, hormone supplements, toxins, and antibiotics. Producers and consumers lined up on opposite sides of the playing field, the cattle industry maintaining a conservative stance while consumers adopted an environmental ideology. One book emerged from the standoff in the early 1990s when Jeremy Rifkin published *Beyond Beef: The Rise and Fall of the Cattle Culture.*

Rifkin, a former Montana cattle rancher himself, echoed many of the same platitudes and concerns that Frances Moore Lappé had earlier set forth. He indicted the cattle culture as "sheer avarice that transformed the great American frontier into a huge cattle breeding ground." He pointed to huge losses to the nation from eating beef: deaths from heart disease, cancer, and strokes; environmental devastation; and "the gross injustice of a world in which the poorest peoples have been starved to support the appetites of a handful of wealthy nations."[32]

Rifkin's book was promoted as "an urgent warning to everyone who cares about the fate of the earth." As Mark Friedberger, an agricultural historian, points out, Rifkin used beef to direct readers' attention to far larger issues: "the wholesale assumption by humanity of industrial values and the effect that these values and the unfettered growth they fostered had on the earth."[33] To Rifkin and Lappé, and many others who call cattle "hoofed locusts," the problem is not cattle but industrialized values that emphasize profit-making. Cattle have become just another commodity from which profits must be wrung. Friedberger believes that the use of the cattle industry, "as a symbol to highlight the spiritual and ecological bankruptcy of the world at large, was misplaced."[34]

Hamburger is embraced by consumers because it looks the least like "red meat" and can be doused with sauces and hidden inside a doughy bun. It is quick to prepare and eat too. Ground meat, made of lower-quality beef, mixed with added fat, sometimes poultry and pork trimmings, and even the addition of wheat or potato flour as filler, is cheaper. As consumers eat more of it, they buy less of cuts like roasts and steaks. Ranchers who raise high-quality meats lose out; their meats are no longer as marketable. Prices have fallen for high-quality, high-bred animals while worn-out dairy cows, whose meat becomes lean hamburger, bring better prices. Incentives for raising high-quality animals have diminished.[35]

Beef itself has changed. "The fear-of-fat people have ruined the beef industry," noted gourmet chef Julia Child recently complained. "Only 3 per cent of our beefsteaks are prime beef anymore. You can't get a marbled steak, even in top restaurants, and the flavor and texture have gone out of the meat. This all started about ten years ago, out of Americans' concern about being so fat." At eighty-seven, the cooking authority observed, "I think it's unhealthy to eliminate things from your diet. Who knows what they have in them that you might need?"[36]

15

The Milk Cow

NOTICE TO THE HELP

THE RULE to be observed in this stable at all times, toward the cattle, young and old, is that of patience and kindness. A man's usefulness in a herd ceases at once when he loses his temper and bestows rough usage. Men must be patient. Cattle are not reasoning beings. Remember that this is the Home of Mothers. Treat each cow as a Mother should be treated. The giving of milk is a function of Motherhood; rough treatment lessens the flow. That injures me as well as the cow. Always keep these ideas in mind in dealing with my cattle.

—Posted in William Dempster Hoard's Wisconsin dairy, 1800s[1]

NO ONE IS SURE when people began milking cows. Archaeologists have found utensils that were used to strain curds and whey dated at 6000 B.C. in Switzerland.[2] Sumerians painted a milking scene in 2500 B.C. that shows a cow being milked inside a corral, with jars of butter and a churn nearby.[3] The milking of cows and the consuming of milk, like the domestication of animals, seems to have arisen in some areas and not yet in others.

Once dairying was established, most cattle were kept not to butcher but for milk or for labor (followed by butchering when they were too old). A milk cow could be kept for years, ate foods indigestible to humans, and provided calves, manure for the garden, and milk. When she grew too old, she too could be butchered and eaten, and her hide used in a variety of ways. Milk, however, unless drunk immediately, has a severe limitation—if not kept cool, it begins to sour and spoil within hours. Some peoples, such as the Celts, were satisfied to drink curdled milk, and the nineteenth-century Irish made a brew that combined milk, cream, beer, and lime juice. Drinking curdled milk was not for everyone, though, and over the centuries a variety of food-processing techniques were developed to utilize fresh milk. It could be made into good things that *did* keep a long time: butter, cheese, curds and whey, yogurt, dried milk, even milk paint, which, when tinted with clay or minerals, covered buildings and furniture with a durable finish that lasted for decades.

Butter is probably the most valuable and most conflict-ridden of the dairy products. Connected with women and secrecy, butter was a mysterious ointment and delicacy that seemed to appear magically, even sexually, when a churn pounded the milkfat exactly right. "Come, butter, come," was a traditional chant, and folklore tells of women who made excellent butter because they did the churning in the nude. Linked to the female and the mysterious, butter has never been an ordinary food.[4] It has always symbolized luxury and was often a sacred food or an offering to a deity. Not all butter was used by wheat-growing cultures as a spread for bread. For example, Tibetans fortify their tea with a dollop of yak butter, which is almost cheeselike.[5]

Roman ideas about butter were shaped early on when Pliny referred to it as the food of the barbarians because it was eaten by Mongols, Celts, and Vikings while Romans relied on oil. Even in

Catholic church ritual, sacred oil is used time and again while "ointments," which describes butter, are believed to be evil and even proof of witchcraft.

Cheese, another milk product, is eaten around the world by a variety of people and seldom causes controversy. Because it stores so well, it has also been valuable as a trade commodity. Pliny the Elder, writing about cheese, noted that Zarathustra, the sixth-century Persian prophet, had acquired eloquence only after living entirely on cheese for 20 years. France has emerged as the premier cheese-making country, today producing more than a million tons each year in more than 350 different varieties. Cheese-making has sometimes been a community affair; in the twelfth century huge wheels of cheese, such as Parmesan, required at least a thousand liters of milk, enlisting the resources of an entire village to create it. The food historian Maguelonne Toussaint-Samat describes how Reblochen cheese earned its name: "It was made secretly, from a second milking, or *rebloche*, and not officially declared by the tenant farmers and shepherds of Savoy. It thus escaped the dues levied by their lords, either lay or religious."[6] In other cases, cheeses such as curds were made in the home and consumed fresh. Cheese was economical as it could be made from the skimmed milk after the cream had been used to make butter, and could be stored for months. In areas where cheeses were not made, the skimmed milk was fed to hogs. The Mongols, who needed lightweight foods because they were nomadic, left skim milk in the sun to dry, making the first instant powdered milk.[7]

While women have been closely linked to dairy cattle because of the economic benefits they provide—food and products to sell—cows eventually became a reliable source for infant feeding, taking the place of mother's milk. As William Dempster Hoard, a nineteenth-century pioneer in Wisconsin dairying stated, "The cow is the foster mother of the human race."[8]

Mothers who were better off had traditionally relied on wet nurses, lactating females who were hired or purchased by a family in order to breast-feed infants. The practice was used in Rome, where wealthy families bought pregnant slave women at the "lacteal" columns in the city market. Families of lesser means hired their wet nurses. Egyptian women usually nursed their own infants, but vases for animal's milk—cattle or goat—created in the shape of a woman feeding her infant, have been found in Egypt.

Wet nurses continued as an alternative to mother's milk into the Middle Ages, particularly among wealthier or noble women who could thus give birth with greater frequency. The child mortality rate dictated that one must have a dozen children to ensure an heir.[9] By the twelfth century wet nurses had formed trade organizations, and the career had become attractive for many young women. It meant status, excellent food, and respect, if not fear, from the family employer. Many aristocratic women sought to become wet nurses for royal babies, which afforded them positions of power and prestige. By the end of the Middle Ages more and more babies of the middle class were being sent away to wet nurses who lived in the country. In 1789, of the 21,000 children born in Paris, 19,000 were sent to wet nurses outside the city. Only the very poor mother, who for some physical reason could not nurse her own child, relied on a bottle of cow's milk for her baby.

Wet nursing was so widespread and popular that problems developed. Mothers sometimes did not see their children until they were weaned, around the age of two, because they were sent such a distance to country wet nurses. Fraud was easy to perpetuate; sometimes an infant was dead for months while the nurse continued to draw money. In other cases old women were suspected of providing babies with cow's milk and chewed-up bread, not nursing them at all. A policeman reported in the eighteenth century that some wet nurses were "drifters lacking husbands, money and morals who turn the nursing of our children into a horrible business, take in seven or eight infants at a time whom they keep alive briefly. . . ."[10]

By 1900 children were no longer being sent to nurses in the country; rather, resident wet nurses were brought to live with a family. It was a safer situation for the infant of the house, but the public began to question the morality of wet nursing because for every woman nursing someone else's child, her own infant was dead (perhaps by infanticide) or wasting away elsewhere. The obstetrician Adolphe Pinard wrote in 1904, "When you see on the public promenades plump and majestic nurses whose heads are ornamented with a bonnet streaming with long and wide multicolored ribbons, carrying a baby in her arms, note that more often than not it means that elsewhere is another poor little one who suffers or is already dead."[11]

Nevertheless orphans, abandoned babies, or syphilitic babies had to rely on a cow, their only chance for survival. Wet nurses resorted to cow's milk at times too, when they could not provide enough milk for their charges. One milkman who traipsed the streets of Paris every morning made his first call "for the wet nurses, to feed the little children." Royal English babies were being given a bottle by the late seventeenth century, being fed "by hand" over wet nurses. Doctors seldom advised it because mortality rates for bottle-fed babies were extremely high; contamination was everywhere and not at all understood.

Bottles for nursing were made from glass, china, or tin; early ones had narrow openings to work as nipples. They were hard to clean and not at all soft and supple—some babies would not suck from them. Nipples were made by clipping off a real cow's teat from an udder (purchased at the butcher's shop) and keeping it in alcohol between feedings until it deteriorated. Cow's teats were recommended by doctors until the 1830s. Cork or leather nipples pierced with holes were later choices.[12] Eventually a glass bottle topped with a long tube became very popular. The bottle could be set beside the baby, who merely sucked at the end of the tube to get the milk flowing. But the inside of the tube was a breeding ground for germs, making the bottles deadly. Rubber nipples that could be attached to any small glass bottle had been marketed since the 1830s, but they were made of toxic materials that included salts of lead, zinc, antimony, and arsenic, added to make the rubber heavier because the nipples were sold by weight.[13]

By the end of the nineteenth century, milk pasteurization had been adopted, and bottles were shaped with simple, wider necks, making them easier to clean. Manufacturers rushed to develop bottles and nipple devices of all types, with brand names like "Perfect Nourisher." What had been an instrument of death became a helpful gadget for mothers. The number of children placed with wet nurses slowly diminished, and by 1907 only 30 to 40 percent of urban European babies were being sent to wet nurses in the country.

Bottles were not the only health hazard of early nursing with cow's milk. The milk itself, in an era before refrigeration, was problematic. Children in the country could avail themselves of fresh, wholesome milk from their family cow, but city babies who relied on cow's milk were given a product that arrived in unclean cans, was full of micro-organisms, and before arriving in the home had been adulterated with a number of substances. Milk was commonly mixed with water to gain profit, then, to make it look whole, additives were mixed in, such as carbonized carrots, grilled onions, caramel, marigold petals, chalk, plaster, white clay, and starch. To replace the cream that had been removed, emulsions of almonds or animal brains were dissolved in the liquid to thicken it.[14]

Pasteurization, in which milk was heated at sufficiently high temperatures to destroy microbes that contaminated it, helped protect against some of the dangers. After 1878 milk was supposed to be boiled before being fed to babies, but it was not always done. Laziness, habit, reluctance to accept advice from authorities, and the stubbornness of some doctors ensured that raw milk continued to be consumed by many babies. Not until after World War II was pasteurized milk more common than raw milk in the U.S. diet. But as milk began appearing in the marketplace in cans or as powder, and advertising promoted its convenience and safety, mothers adopted it and abandoned breast-feeding, particularly when admonished that "the woman is not a pantry."[15] Feeding babies with refined manufactured products seemed the correct thing to do as the consumer society dawned.

DAIRY WORK, whether for the family-owned cow, a small herd, or as a means of employment for someone else, historically has provided one of the few trades in which women and men could compete on an equal footing. Young farm girls in Europe and North America sought positions with neighboring farm families, where they learned skills at dairying that would later benefit them when they married and had their own cow or small herd. Dairying was hard work and included much more than just milking the cows. Dairy maids were responsible for scouring and washing all the pans and pots, and for salting, smoking, drying, and storing cheeses. They also processed grease, did weeding and haymaking, and hoed in nearby fields when other work was done. In dry areas, much of their day was spent carrying buckets of water to thirsty cows.

Backbreaking labor was part of the labor-intensive dairying industry. But it was worth it. In eighteenth-century Tuscany, for example, a girl's prestige was enhanced by the dairy farm she worked at. A skilled dairy maid could command higher wages and more respect because a farmer could profit from her preparation of cheese and butter. And those who learned a range of skills were more marriageable.

Between 1500 and 1800 there were no changes in dairy technology, but it was not the cozy pastoral scene most romantic painters portrayed. A dairy maid was known by her "red plump arms and hands and clumsy fingers," and the cows were usually milked in the field, any time of year, which meant frost and snow and freezing fingers while cow and girl stood in mud. Dairy maids worked from the first milking at 4 a.m. of two dozen or more cows, then prepared the milk and cream: separating it, scalding and scouring the pans, making butter and cheese, and performing every other task that could fill the day. At 4 p.m. they tramped back to the fields with buckets for the evening milking. Dairy maids who were able to work in stall-fed cattle barns did not have to go out to the fields, but on the other hand they were forced to remain in a sweltering, fly-ridden, stench-filled barn, which they had to shovel and sweep clean.

Although milking and dairying were important jobs to the employer, women's wages never equaled men's. "What it did do was to establish a relationship between women and the exploitation of the cow which was very important," the historian Olwen Hufton points out. "For the most marginal of farming families, the cow, where grazing could be obtained, represented the best source of fat and protein for home consumption and ideally for wider sale. In most countries it was perhaps the first thing bought from a bride's assets on marriage. In Ireland dowries were counted in cows. Skill in dairying was therefore the single most valuable talent a farmer's wife could command." In Britain, dairy maids could apply at better farms when jobs opened up, or take themselves to hiring fairs, usually held every Martinmas (November 11). There prospective employers and workers could bargain with each other, followed by a holiday celebration. The dairying regions of Britain, the Netherlands, Flanders, and northern and western France were areas where female labor was sought; elsewhere there were few opportunities for women to find employment.[16]

The American dairy industry grew slowly from its early center in New York state. Without refrigeration or transportation, dairies were small local ventures; most farmers found raising grain more profitable. Cheese-making was rude: the milking was done in an open yard, then the curds were worked in wooden tubs and pressed into hollowed-out logs. "Everything was done by guess, and there was no order, no system, and no science in conducting operations," an 1865 USDA report noted. Quality was low, making it difficult to create an export market.

In 1851 a farmer in Rome, New York, changed the way cheese-making, and hence dairying, was done. Jesse Williams was a successful cheese maker who had no trouble selling his high-quality product. When his son married and began dairying, Williams signed a contract for the cheese made on both farms at seven cents a pound—a very good price at that time. But the younger Williams could not produce the same quality cheese as his father, and the distance between the two farms made it difficult for the son to work his own cows, then travel to his father's dairy for instructional lessons on cheese-making. So the elder Williams hit on the idea of combining the milk from both dairies at his farm and making all the cheese at the one location. Instead of the cheese maker traveling between farms, the milk was sent by wagon each day. The system worked so smoothly that the Williamses eventually began buying milk from adjoining dairies and built additional processing facilities. It was the beginning of what became known as "the associated dairy system," something entirely new, and in Europe called the "American system of dairying." By the time of the Civil War the associated dairy system had been established across the Northern states and into Canada.[17]

It was the advent of railroad lines that really inaugurated an American dairy industry. In the 1870s farmers in Rutland, Vermont, 241 miles from New York City, were shipping milk to the city by rail. Customers in St. Louis were receiving milk by rail from dairies up to 95 miles distant. Railroads created a much larger market and made it financially feasible to raise dairy cows rather than meat livestock or grain crops. Rail transport also opened the milk industry to widespread tampering because the milk passed through so many hands on its way to the consumer. Each person who received the milk diluted it with water to gain a larger profit; by the time it reached the dinner table it had been weakened and doctored with coloring and thickening agents to disguise the fact. The watering-down of milk hurt the farmers as much as the consumer, because their customers ended up drinking a lot of high-priced water instead of whole milk.

DAIRY COWS played a large role in the settling of the American West. In particular, the presence of American cows helped shift the Pacific Northwest from British possession to become part of the United States in the 1840s. American women, coming from a northern European dairy heritage, refused to move west without a family cow. In the early 1800s this reliance on a milk cow was readily understood by leading executives of one of the world's largest corporations, the Hudson's Bay Company. Hudson's controlled the fur trade in Canada as well as most of North America west of the Mississippi River at the time and was adamantly opposed to allowing either white women or private cows to enter their domain. Indian women were vital as trappers' wives and workers. They prepared hides and did a good deal of trapping as well as being a market for trade goods. White women coming west meant farms and settlement—and that would ruin the fur trade. Since Euro-American women would not settle on farmsteads without a cow, Hudson's policy of refusing to sell a cow to a settler kept farmers and their families out of the Northwest.

Company officials were shocked in the summer of 1836 when word spread that two American women had ridden across the continent by sidesaddle and had arrived in the Pacific Northwest. The first two American women to enter the region, Narcissa Whitman and Eliza Spalding, had gone west with their missionary husbands and brought along a small band of cattle. The animals they brought were not Texas-type longhorns but small, multi-purpose breeds useful for meat and milk. Other emigrants quickly followed. Families in wagons pulled by oxen and trailing small bands of cattle began trekking west on the Oregon Trail. The Hudson's Bay Company's hold—and Britain's—on the West was broken.

Narcissa Whitman, on the journey across the continent in 1836, wrote from their camp "West of the Rocky Mountains": "Our cattle endure the journey remarkably well. They are a source of great comfort to us in this land of scarcity. They supply us with sufficient milk for our tea & coffee which is indeed a luxury. We are obliged to shoe some of them on account of sore feet."[18] As the hot, dry trek wore on, Narcissa wrote that she and husband Marcus enjoyed reminiscing about home, particularly envisioning his mother—"thought a sight of her in her Dairy would be particularly pleasant."[19] Crossing the Portneuf River was difficult; it was the widest river crossing the party had to make on horseback, and the cattle were plagued by mosquitoes. "It seemed the cows would run mad for the Musquetoes we could scarcely get them along," Narcissa noted. By the time they reached the Snake River, in what is now Idaho, she added, "We think it remarkable that our cattle should endure the journey as well as they do. We have two sucking calves that appear to be in very good spirits, they suffer some from sore feet, otherwise they have come along very well."[20] The Whitman-Spalding party had taken seventeen cattle from Missouri, the first cattle to be driven over the Rockies through to Oregon. Only eight made it to Fort Walla Walla. When Indians met the missionaries at the Rocky Mountain fur trade rendezvous in 1836, the natives were interested in the women's clothing, the wagon or "land canoe," and the cattle.[21]

When the missionaries reached Fort Vancouver they were surprised to discover that the Hudson's Bay Company owned so many cattle in the Pacific Northwest—"1000 head in all their settlements." Fields of turnips were "large and fine," providing excellent winter fodder for the cattle. Narcissa visited the large dairy there and estimated that between fifty and sixty cows were milked daily.[22]

When U.S. army Lieutenant Charles Wilkes led an expedition to explore the Pacific Northwest, he visited the Walkers' Protestant mission, not far from the British trade post at Fort Colville. The men "all passed me as I was milking," Mary Walker noted with chagrin. One of the expeditionary members told her that "the most pleasant sight he had seen in Oregon was a lady milking her cows."[23]

The mission women were like other rural women worldwide who relied on their cows for the basics of family survival: they used rennet (from deer stomachs) to make cheese from skim milk, churned butter, and eventually "milked six cows morning and night." Mary wrote one winter that she "dipped twenty-six dozen candles" made of "very white and nice" beef tallow. She wrote about preserving the meat: "salted a keg of beef by the rule four qts. salt, four lbs. sugar and four oz salt peter to 100 lbs." Without fencing, the cattle roamed free, and in order to milk, Mary caught the calf, which made the cow come near for nursing. As the calf nursed, Mary milked the other teats. When it was difficult to catch the calf, milking took more effort than usual.[24]

Another Oregon missionary, the Reverend Jason Lee, a Methodist stationed in the Willamette Valley, had driven two cows from Independence, Missouri, to the Oregon Country in 1834.

Shortly after, his fiancée arrived by ship and married Lee, whereupon he was determined to bring privately owned cattle into the valley.

At a meeting in his mission house, Lee organized a joint stock company to raise money to send men to California to obtain Mexican cattle for American settlers in Oregon. Ewing Young, a seasoned mountain man who had been in Spanish California, was named to head the party. In January 1837 eleven Americans and three hired Indian hands were given free passage on an American brig, the *Loriot*, to California, to drive the cattle north to Oregon.

The new Mrs. Lee wrote to her brother back east in 1837, "Beef is scarce, and all the cattle that the settlers here have used belong to the fort. They would not sell, but lend as many as any person wishes to use." She related the expedition to California and claimed, "we will have eighty head of cattle—we will have plenty of milk and butter in the future. . . . We cannot make soap on account of not having fat and have been obliged to pay fifteen cents a pound at the fort."[25] Tallow and fat were difficult to obtain locally because no one voluntarily slaughtered an animal. Tallow was imported from Mexican ranches in California, brought north by sea.

In California Lee's party bought eight hundred head of cattle from Spanish government officials at three dollars a head. The Americans then began a very difficult overland drive, heading the animals north toward Oregon in the West's first overland trail drive. The animals were near wild and had never been driven before; just crossing the San Joaquin River was a monumental task. One herder wrote, "Another month like the last, God avert! Who can describe it?"[26] The animals had to be pushed and pulled through the underbrush and over the mountains at Mount Shasta, and the drovers' tempers flared. Despite rough terrain, lack of cattle-driving skills, and a few ambushes by Indians, they reached the Columbia River with 630 animals in mid-October, after eighteen weeks on the trail.

That band of cattle laid the foundations for Oregon's livestock and dairy industry, and eventually the Pacific Northwest's independence from Britain. "The effect on morale was tremendous," writes the historian David Lavender in *Land of Giants*. "Until this moment the thirty-odd settlers in the Willamette had possessed almost nothing they could call their own. The titles to their land were clouded, since no sovereignty yet existed. Their crops had value only if the Hudson's Bay Company chose to buy. These stringy, fence-breaking cattle, however, were something on which a man could stamp his brand and say, 'This is mine!' As such, [the] epic drive had a significance far transcending its size or its monetary value; for it was the Northwest's first gleam of independence, the first crack in the Bay Company's benign but hitherto invulnerable monopoly."[27]

Eventually the cattle situation in the Columbia River region changed; by 1846 the fur trade had declined, the Hudson's Bay Company had moved their operations north into Canada, and the region was no longer in the economic grip of British interests. The settlers who had come into the country, most trailing milk cows and oxen along the Oregon Trail, meant to stay. The cattle population, husbanded so carefully, soon burgeoned, and by the 1860s cattle were being driven east to gold-mining regions in Idaho and Nevada.

In the 1870s, when the buffalo were nearly gone, the grassy plains of Montana and Wyoming beckoned with potential wealth for cattle grazers. But where were the animals to come from? The Wyoming Cattlemen's Association refused to let in tick-ridden Texas cattle. In the Pacific Northwest an abundance of animals were made up of New England and Midwestern breeds that had been trailed west. They were superior animals to the rangy Texas longhorns, but how could they be brought to Wyoming's grass?

Several small herds made the overland trek before the *Cheyenne Sun* recognized the news potential of the movement. In 1879 it heralded G. A. Searight's proposed cattle drive as "the greatest cattle driving enterprise of the age." Searight planned to take a group of Texas cowboys westward by train to northeastern Oregon, where he planned to buy 5,000 cattle. When Searight's 27 cowboys left Cheyenne, a brass band at the railroad station sent them off. As if the men were off on a mission to the moon, the *Sun* explained,

"The expedition which left Cheyenne today was composed almost exclusively of Texans—young men who have spent years on the trail between Texas and Kansas. They regard this trip as equal to two trips to Texas, and as many of them will be absent on the trail nearly two years their farewells to friends were of a serious nature." The drive was successful, and others followed. In 1882 another Cheyenne newspaper reported that 100,000 animals were on the trail from Idaho, Oregon, and Nevada eastward. The drives were part of what has been overlooked in the mythology of the cattle-drive era of the West: pointing them east.[28]

DURING THE EARLY 1800s American farm families had kept a cow or two, the care and production of dairy products was the responsibility of the women of the household, and methods were crude. Pans of milk sat on the doorstep or inside the kitchen, where the cream was allowed to rise before being skimmed off and used for butter. If possible, a family built a stone spring house where milk was set in crocks or pots allowing cool water to flow around it. The quality of butter and cheese varied from house to house; it was usually unsanitary if not rancid. On many farms it was held all year until enough had accumulated to take it to market. Low prices reflected low quality. The Civil War created a market for cheese, and in 1865 prices rose to more than twenty cents a pound. The market created opportunities, and cheese factories appeared in Pennsylvania, Ohio, and other states; by 1869 there were more than a thousand factories, and cheese-making had moved off the farm.[29]

Factory processing of milk and cream was successful, but there remained a need for some way to preserve pure milk. In 1846 in New York, Gail Borden began experimenting with condensed milk and eventually perfected a high-quality canned condensed milk that swept the marketplace. By 1889, 38 million pounds of condensed milk were being produced annually.[30]

In an era of mechanical invention, the dairy industry provided many opportunities for creative thinkers. For a time, forty to fifty new butter churns were patented each year, and a new churn was patented every ten or twelve days for more than seventy years. No industry lent itself to such ingenuity. Creamers, butter workers, churns, agitators, milking machines—the list of inventions, and the potential for more, was endless.[31]

The mechanical cream separator was probably the most useful piece of equipment developed during this time. A European invention, it was immediately adopted by almost all dairy operations. It replaced much of the labor involved with separating cream from milk by using centrifugal force, possible because the specific gravity of milk is greater than that of cream. The equipment could be purchased in various sizes and powdered by hand or a treadmill run by a dog, sheep, horse, child, water, electricity, or steam. With the mechanical separator, milk and cream could be separated on the farm, the cream sent to town to the creamery and the skimmed milk kept for family use or more likely fed to hogs. The mechanical separator was the first step in taking milk processing out of the home, as pans of milk no longer needed to be set about the house waiting for cream to rise so that it could be skimmed off by hand. It was the first step, too, in moving dairy processing out of female hands. With the rise of mechanical equipment on the dairy farm—separators and later milking machines—women were slowly removed from responsibility. As dairying took on industrial overtones, women's place was superseded by equipment.

Cows were "reinvented" too, using breeding and feeding techniques. When dairying in the United States began, a cow that would produce a pound of butter a day for two or three months was a local celebrity. One cow, the "Oakes cow" became famous in Massachusetts in 1816 because she gave 44 pounds of milk a day and made 467 pounds of butter in one season. By 1899 a cow was expected to produce 5,000 pounds of milk a year. At the end of the nineteenth century the milk of a single dairy cow made as much butter as that from three or four average cows at mid-century. Careful selection for breeding, and culling of animals with low productivity, along with better care and feeding had promoted higher productivity.[32]

Cows were put under additional scrutiny when the Babcock test for fat in milk was developed. The result of research at an agricultural experiment station in Wisconsin, the Babcock machine used centrifugal force and chemical action to sample milk. It was simple and inexpensive and could test from two to forty samples at once in just a few moments. Sulfuric acid and water were needed in the testing process, but "any person of intelligence" could learn to use it, and it became a significant economic advance in dairying. With the Babcock test, milk could be assessed for fat content and priced accordingly. No longer could milk be watered down and sold at full value. The Babcock test also led to ranking cows according to their production. In addition to a plentiful supply of milk, it had to have a high fat content too. Interestingly, Dr. S. M. Babcock, the device's inventor, refused to patent it, making it available to all. Such generosity in an era infatuated with invention and patents was recognized by a medal from the Wisconsin legislature and an award from the dairymen of New Zealand.[33]

By 1899 the cow was viewed as "almost as much a machine as a natural product, and as already shown, a very different creature from the average animal of the olden time." The factory system had replaced home dairying so completely that one agricultural writer noted with pride, "The cheese vat or press is as rare as the handloom, and in many counties it is as hard to find a farm churn as a spinning wheel." The industrial age had enveloped the farm, but in the United States there was still one milk cow for every four persons.[34]

Industrialism and science pushed the dairy industry in new directions no one could have predicted. Average milk production per cow eventually soared due to improved feeding practices, selective breeding, and in some cases hormone and antibiotic supplementation. Today the average amount of milk given by a dairy cow has risen to about 17,000 pounds during its ten-month lactation period. The record holder is a Holstein in Marathon, Wisconsin, named Muranda Oscar Lucinda-ET, who produces 67,914 pounds per year. Such cows are valuable not so much for their milk but for their eggs, which are harvested, then artificially inseminated and kept as frozen embryos in laboratories and shipped around the world for insertion in less-productive cows.[35]

Almost every scientific or technological innovation in dairy science has only made the industry more competitive. As more techniques were devised to raise the amount of milk produced per cow, the price farmers received for milk dropped. The U.S. market has been swimming in excess milk for years, but government and industry research continues to seek ways to get more from a single animal, turning a cow into a factory in which production can be continually increased. The result is an industry in which the number of dairy farms continues to decline in a steady trend that began early in the twentieth century. In Washington state's Snohomish County, for example, where four hundred dairies thrived in the 1960s, today there are fewer than seventy-five. The shrinking number of farms is attributed to several factors: increased feed and labor costs; environmental regulations against manure and urine contamination of groundwater, which requires expensive containment systems; encroaching urbanization in some areas; and for farmers with no ability to expand beyond a couple of hundred animals, no way to compete with big operators.[36]

Dairy farms have been too productive for their own good. Even with a shrinking number of farms, America's 9.2 million cows still provide 18 billion gallons of milk each year, more than enough for American demand. California's dairies make it the nation's largest dairy producer: a million cows generate $4 billion worth of milk and cream—twice as much as the earnings from grapes, the state's number two crop.[37] Only about a third of all fluid milk is sold as a beverage, the rest is processed into ice cream, butter, yogurt, sour cream, cheese, and dried products like whey protein concentrate. Dairy products make up more than 10 percent of grocery store food sales. The jobs related to the dairy industry don't stop at the farm; they also include geneticists, processing-plant workers, veterinarians, government inspectors, all the way down to grocery-shelf stockers and the creative types who design clever packaging and advertising campaigns.

Today milk is pasteurized, chilled, and conveyed in sealed systems from the time it leaves the cow's teat via milking machines to when the carton is opened by the consumer. But in spite of the low cost of milk and its sanitary preparation, American consumers are drinking less each year. Why? School lunch programs still feature milk, but many also make juices available, and scarcely a public school in the country does not allow vending machines with soda pop, which children eagerly consume because of the sugar and caffeine. Diet-conscious Americans have limited their consumption of milk and dairy products in efforts to lose weight. Young women hesitate to consume the calories in milk when diet sodas are available.

Health issues have also developed over the dairy industry's use of antibiotics, in the case of dairy cows to prevent infection and disease because animals are kept in close quarters. Antibiotics are not used in the dairy industry to increase the weight gain of cattle, as they are for beef producers, because all of a cow's energy must make milk, not meat. The heavy prophylactic use of antibiotics to maintain the health of cows is of concern to consumers who worry about creating drug-resistant bacteria from antibiotic overuse. And overuse is prevalent because intensive practices such as the use of hormone supplements to increase milk production lead to increased incidences of infections in cattle. Their overworked bodies are susceptible to bacteria and viral infections, and their greater numbers (thousands of cows in a confined dairy) increases the sprcad of contagious disease.

Consumers' greatest fear, however, comes because a third of the industry uses hormone supplements to increase the amount of milk each cow gives. A synthetic growth hormone, bovine somatotropin (manufactured by Monsanto as Posilac), has been approved by the Food and Drug Administration for dairy cows. It is similar to the hormone that makes baby calves grow so quickly and is given to cows because it causes them to produce more milk. The FDA claims that milk and meat from treated animals is safe. Consumers are not convinced, and after past scandals over the use of hormones, such as the diethylstilbestrol (DES) disaster (which caused cancer in daughters of women treated with a similar hormone), they would prefer natural, untreated cows. They are unable to make the choice in the marketplace, however, because FDA rules essentially prevent milk producers from labeling their product "hormone free." The FDA believes that would cast a cloud of suspicion over milk not so labeled. At this writing there is no way to tell if milk has come from hormone-treated cows. Even dairy laboratory tests are inconclusive.

There has also been a growing anti-milk movement based on animal rights activism, which points to the short life span of a dairy cow and the cruel conditions enforced by industrial dairying. Dairy cows first begin lactating at two years of age; they have a working life of about four years, then, as their production declines with age, they are sold to meatpackers. Because they are bred for milk and not meat, their meat is tough, so their carcass is usually made into ground beef.[38] Male calves are slaughtered while young because they have little value as beef animals, and they will never provide milk. The calf's stomach is valuable because it contains rennin, an enzyme which is made into rennet, which causes milk to curdle and is used in cheese-making. The calf's skin is also sold to the leather industry, where it becomes handbags, belts, gloves, and fine shoes.

PETA, People for the Ethical Treatment of Animals, an animal rights activist group, is adamantly opposed to the dairy industry. The Physicians Committee for Responsible Medicine also advises against drinking milk. The anti-milk movement has become almost a cottage industry through the efforts of activists on the Internet. Some, like Robert Cohen, author of *Milk, the Deadly Poison*, have financial interests in marketing soy milk as an alternative. Soy milk is a new technological development but does not replicate cow's milk. Cow's milk contains every essential nutrient needed by humans; soy milk does not. And soy contains large amounts of estrogen, a boon for middle-aged women but not so beneficial for males.

Health advisers have been telling consumers for years to be sure to consume plenty of calcium; it has become almost a mantra for women seeking to avoid osteoporosis. Milk provides a reliable source of readily absorbed calcium, but processors have begun adding additional calcium to make milk even more nutritional. The extra calcium is usually added in the form of calcium carbonate and carrageenan. It recalls the nineteenth-century days of milk adulteration, because calcium carbonate is simply limestone or chalk, which was one of the most widely used adulterants before the 1906 Food and Drug Law forbid the watering and adulterating of milk. Today what was called an illegal adulterant in Teddy Roosevelt's time is termed a fortification and a reason for additional price points in the marketplace. Carrageenan, a seaweed form of algae, is added to rid the milk of a chalky-gritty taste and to make it feel creamy in one's mouth. It is also added to cheaper ice creams to give a creamy taste in place of real cream. Other types of calcium fortification come from calcium gluconate, a fermented corn sugar which is treated with calcium carbonate.

While regular milk is being fortified with additional calcium, additive research continues, producing skim milk that tastes thick and rich because carrageenan, cellulose, and artificial colors and flavors are added. In California, milks are commonly calcium-fortified, bringing higher prices in the marketplace and pushing out products from out-of-state companies that do not add calcium. In addition to calcium, "heart-healthy" milks are being formulated with hydrolyzed oat flour added to improve the texture of fat-free milk and to convey the putative benefits of oat fiber in fighting chronic heart disease. Eventually ice creams with added oat fiber will be marketed as "good for your heart." These additions are all good for profits too, because they ultimately replace high-quality milks and creams with cheaper additives.[39]

All of this "new technology" brings higher prices because it enhances the nutritive value of the milk. Or does it? Nutritionists point out that no matter what the source of calcium intake, only about 30 percent of it is actually used by the body. Other minerals are required for calcium absorption, in particular magnesium. And the kinds of calcium additives being used in milk are questioned by some who maintain that dairy products provide usable calcium while other sources are not as bioavailable—the body cannot utilize them as well. So even calcium-fortified milks may not provide adequate amounts of calcium because it may not be usable by the body. Gritty calcium salts that require carrageenan to keep them in suspension are not optimal nutritional sources of calcium. And just as dairy farmers lost market share to middlemen who watered down fresh milk, adding calcium to milk means consumers need to drink *less*, actually shrinking the market for milk.

Ninety-eight percent of American households purchase milk, and the average American drinks about twenty-five gallons of it each year. But despite population increases the nation as a whole does not consume any more milk than it did in 1974. Americans have switched to other beverages. If consumers continue to run from the product, fearful that it contains antibiotics, pesticides, hormones, and bacterial contamination, it's highly unlikely that any amount of technological innovation will expand the market.

Consumers *do* want healthy, natural milk and dairy products, yet the industry seems to ignore the organic market, focusing instead on larger and larger production entities. Western states like Idaho and Montana, which once had small dairies and not much population, have lax agricultural and environmental laws and now are becoming home to mega-dairies of thousands of cows, shipping milk to several states in the beginning of a trend that promises a variety of problems. Isolated from population centers, such mega-dairies and their huge amounts of waste, pesticide use, and environmental problems can be obscured. While the environmental problems are most immediate, the oversupply of milk ultimately is a greater concern. Until demand is increased, or even holds steady, there is no need to push dairy production in the United States.

There *is* another vision for the future. New technology imported from Israel's kibbutz farms may ignite opportunities for small dairy farmers. A system called the Pladot Mini Dairy Milk Processing System, developed on a collective farm in the 1980s, makes it possible for farms of thirty to forty cows to process their own milk and make cheese, yogurt, butter, and other dairy products which they can market to consumers directly. The systems sell for about $85,000 and promise a new kind of dairy enterprise—small, less reliant on chemicals and drugs, and more environmentally friendly. That number of cows, considered far too small for a successful dairy operation in most areas of the United States, would enable the farmer to keep healthier cows, create wholesome products, and spread manure on pastures, changing dairy pollution to a benefit. Niche marketing might even bring back door-to-door milk delivery, something that disappeared from U.S. doorsteps in the 1980s but is still quite common in Britain.[40]

Home delivery, something the e-marketers have heralded as an Internet shopping advantage, was pioneered by the dairy industry long ago. It is impossible to return to the days when a milk cow was led door to door for milking, so that consumers could see they were getting the real thing, fresh and unadulterated—but wouldn't *that* be a marketing gimmick!

16

Ruminations: Care and Feeding

Principles have no real force except
when one is well fed.
—Mark Twain

WHEREVER THERE IS TALK of cattle, the argument soon surfaces that we should not raise livestock at all because they eat plants that would be better utilized by being fed directly to people. Beef is viewed as an elitist foodstuff, something modern technology and ideology has moved beyond. But in fact cattle are low on the food chain, eating grasses and plants that are not digestible by humans. Cattle do not compete with humans for food unless they are purposely fattened on grain, which is an unnatural feeding situation (though highly profitable). Cattle are naturally meant to eat grasses, plants, and tender buds and shoots. They lack upper incisors, so they cannot bite through tough plant material. Their tongue sweeps the plant material into their mouth, where it is cut off by the lower teeth and swallowed. It enters the microbe-laden rumen where it is partially digested. A few hours later, as the cow rests, it burps up the rumen contents, grinds it between the molars, and reswallows.

Cattle are *ruminants*, a term used to describe animals that have split hooves, eat plant materials, and chew their cud. Their complicated digestive system allows them to digest foods in stages, meaning they can eat high-cellulose foods that other animals cannot. Ruminants include rabbits, goats, sheep, deer, llamas, antelopes, and, of course, cattle. What is fascinating about ruminants, particularly to agricultural scientists, is the idea that the rumen makes nearly anything digestible. The rumen actually creates its own food nutrients because it synthesizes essential amino acids and vitamins from nitrogen-containing compounds through microbial action. This allows cattle to convert inferior proteins and even nonprotein nitrogen-containing compounds (such as urea) into superior protein in milk and meat. Cattle actually create their own B-complex vitamins, something entirely unique to ruminants.

In the mid-twentieth century, cattle research moved away from the idea of looking for an optimal feed for the animals' health; instead it began to concentrate on using worthless, indigestible waste material as feed. The rumen, with its zillions of busy little microbes, could be used to recycle industrial wastes, turning a profit for industries who could sell materials they would otherwise have to pay to dispose of. Even more efficient than the early Chicago meatpackers who bragged they sold every part of the pig except the squeal, livestock feedlots could serve as society's garbage dumps. Worthless waste could be sent down the throats of beef and dairy cattle as feed, and farmers and ranchers would even pay for the products. The rendering industry, at one time a supplier of glues and soap, boomed into a large-scale, international livestock feed source.

What happens in a cow's rumen began to be investigated in 1928 when a North Dakota agricultural experiment station researcher published the first article describing fistula surgery in cattle. It was easy enough to do: a window or opening was cut in the side of the cow's abdomen so that the rumen could be observed and sampled without killing the animal. Tubes or windows were installed in the cow's side, and researchers began testing rumen temperatures and responses to various liquid or solid materials by retrieving samples of feed from the rumen at various stages of digestion. "Fistulating animals apparently causes them very little discomfort or difficulty," an animal science textbook claimed. "They live a quite normal life and provide a ready source of micro-organisms for laboratory studies."[1] What came of the fistulated cattle studies, besides many research grants and numerous well-funded careers for agricultural researchers, was the idea that nearly anything could be put into the animal's digestive system. By the 1960s fistulated studies had led researchers to try using lignin, corncobs, and even wastepaper as cattle feed.

Agricultural researchers in the 1960s, reeling in cold war rhetoric and research funding, sought to raise the prestige and importance of their work by incorporating radioactive materials into livestock research in an attempt to compete with the space program and the budding computer industry. "Future animal scientists . . . will probably rely more on techniques using radioactive isotopes than on any other method known today. Radioactive tracers and radiation sources have become indispensable to most phases of agricultural research," a textbook advised. Fistulated stomachs seemed mundane when agricultural science textbooks began including photographs of nuclear reactors as future research components. Research on radioactive isotopes did make possible one line of investigation in cattle: the use of hormone supplementation. With radioactive studies, hormones could be fed to cattle and tested to show what level of hormone remained in the meat.[2]

Cows were players in other cold war research too. Fearing Soviet missile attacks, researchers studied whether cows could act as radiation shields for humans. Sparsely populated regions in the American West, particularly grazing lands, were testing grounds for radioactive fallout. The U.S. tests scattered radioactive fallout across the region, where it fell on crops and pastures. Cattle naturally came into the picture. Along with grass and soil, they too were sampled for radioactivity. In the 1960s researchers found that 82 percent of the strontium-90 contamination in Western residents came from eating field crops, 17 percent from drinking milk, and a negligible amount from meat. Field crops directly contaminated humans, but cattle appeared to act as a filter of sorts; if they ate contaminated grass, it appeared that somehow they eliminated the radioactive isotopes. People were also advised to drink lots of milk, as calcium would displace strontium in the human body. Fortunately Americans did not have to worry about contamination of the meat and milk supply by Soviet attack, and the research projects quietly faded away.[3]

TODAY cattle feeding is a huge industry, heavily invested in research. What began with the adoption of clover and turnips as cattle feed in the eighteenth century has moved to reusable plastic pellets, fed along with liquid nutritional supplements. The pellets, extracted from the carcass at slaughter, can be fed again and again. Today's commercial cattle, whether dairy or beef, are being fed unusual combinations of feedstuffs that are not only surprising but in many cases problematic and even disgusting.[4]

Beef cattle nutrition textbooks list the most appropriate feeds for both weight gain and least expense. "Wheat flour which had become damp in railroad cars, and thus unfit for human consumption, is an excellent cattle feed when mixed into a slurry. Peanut butter, stale cake mixes, stale bread, broken cookies, stale candy bars, and almost any material which (1) is reasonably digestible by cattle, (2) does not contain toxic materials, (3) is not extremely unpalatable, (4) does not contain too much crude fiber, and (5) is economical" can be used in cattle feeding. While stale bakery items are innocuous, the textbook urges that "any different or unique feedstuff which meets the five qualifications indicated probably should be investigated and evaluated as a potential cattle feed."[5]

Different or unique, certainly, because the textbook goes on to point out that protein feeds for cattle can be made of urea; blood meal from slaughterhouse waste; meat and bone meal from mammalian hair, hoof, horn, and hide trimmings; fish meal; feather meal (ground-up poultry feathers); dehydrated poultry waste (said to be equal to good-quality hay); dried chicken manure; and—the ultimate recycler's dream—dried cattle manure fed back as "wastelage." When feeding statistics for cattle fed a diet of traditional corn, cottonseed meal, and alfalfa were compared to cattle fed a similar diet that included 40 percent manure, the weight gain was similar. When the cattle were fed "wastelage" (a combination of manure and silage made of cornstalks and straw), their weight gain was *greater* than the gain of those fed traditional cattle diets. No need to wonder where all the cattle manure is going—it is definitely innovative and environmental to put it back into the cattle, and ultimately into the human diet. Of course old standbys are still being fed to cattle—cottonseed meal, soybeans, canola meal, flaxseed meal, corn gluten, distiller's grains, and brewer's grains. But high profits drive business decisions; the lure of finding cheap feeds that the rumen can digest continues to spur agricultural research.[6]

What is problematic about cattle feeding today is that agricultural research and human public health research seldom intersect. Feeding animal excrement to livestock is an example of an area that has escaped public health scrutiny. As the Physicians Committee for Responsible Medicine noted in *Preventive Medicine*, foodborne illness is a common and serious problem, and despite efforts to improve slaughterhouse procedures and warn consumers about food preparation, little attention has been paid to animal waste used as livestock feed. In 1994, 18 percent of the poultry producers in Arkansas fed a thousand tons of poultry litter to cattle, and the practice is common in other areas of the country as well. Is this the best way to eliminate the 1.6 million tons of livestock wastes produced in the United States each year? Consumers may gag at the idea of excrement in livestock feed, the industry answers, but how many bags of manure can the neighborhood garden store sell every spring? There has been literally no outcry in the press, no public demonstrations by environmental groups. Manure disappears into animal feed and is returned in a different, cattle-palatable form to the same feedlots where it originated.[7]

What alarms the Physicians Committee, and should give the rest of us pause, is that manure-based feeds may not be processed at high enough temperatures to kill pathogenic salmonella and *e. Coli* microbes. Another concern the doctors voiced is that using animal wastes as feed presents the possibility that antibiotic-resistant bacteria may spread from one animal to another. As of 1997, when the Physicians Committee article appeared in *Preventive Medicine*, "few research reports have addressed the safety of this practice, and . . . further microbiological studies are recommended to assess the extent of risk."[8]

Recently European farmers were horrified to discover that their livestock feed had been tainted with dioxin and in France included human sewage. In what is an elusive and closely guarded industry, with few but large commercial interests, the animal feed industry provides a way to free the public from worrying about how to dispose of the most distasteful of industrial wastes: sewage, manure, offal, animal by-products. But unless the industry changes its practices, feed bans will become more common in the face of outbreaks of contamination and links to diseases passed through feeds, such as Mad Cow disease and foot-and-mouth disease.

WHILE cattle feed lot practices are suspect, grazing also has its critics. Cattle grazing on public lands is one of the most incendiary issues in the Western United States today. Environmentalists and ranchers have been urging a showdown in the media and in the voting booths of Western states for decades, and the issue has not faded away. What is the fuss all about? Just a few large mammals, crunching grass on marginal land? In fact it goes far beyond the animals and beyond grasslands. The issue is political and divides communities and even families over the question of grazing as good or evil.

Outspoken anti-ranching advocate Lynn Jacobs, in his book *Waste of the West: Public Lands Ranching*, epitomizes grazing's critics. Jacobs wrote the book after years of hiking and camping on Western public lands where he saw devastation wrought by overgrazing.

"Ranching has wasted and is wasting the western United States more than any other human endeavor," he writes.[9] His goal is to end grazing permits that are sold by agencies of the federal government (the Forest Service and the Bureau of Land Management, among others), which allow private cattle owners to graze their animals on public lands. He points to the heyday of Western cattle ranching in the 1880s, when 35 to 40 million cattle were put out onto the plains to graze, as "forage fever, similar to gold fever, swept the West."[10] When bad weather devastated the overstocked ranges, hundreds of thousands of dead cattle were strewn across the plains. Although the numbers of cattle fell after that, the number of sheep increased to 53 million by 1890. The grass was irresistible to Gilded Age capitalists.

Jacobs describes the class system that developed on Western rangeland, with aristocratic cattle "barons" shaping laws and lawmakers while forcing out small farmers, family herdsmen, and sheepmen. "More western homesteaders' hopes and hard work were crushed by ranchers and their cattle than by any other influence," he writes. Competition and violence erupted into early range wars as cattlemen fought over grassland and water.[11] It was a difficult, lawless time, similar to the mining frontier. Without access to law enforcement and adequate judicial authorities, the early years of the Western territories were indeed filled with fraud, violence, and lawlessness.

In 1891 the U.S. government set aside federal forest reserves on public lands. The U.S. Forest Service was established in 1905, and cow and sheep grazing permits were sold to pay administrative costs. This helped control the land, but essentially on paper. The public lands made up nearly half the seventeen Western states, and there was little oversight of their management. Grazing continued with no restrictions until it was clear that overgrazing and abuse needed to be brought under control. In 1934, Congressman Edward Taylor, a Colorado rancher, introduced the Taylor Grazing Act, which was adopted expressly to eliminate the nomadic herding that was blamed for wreaking havoc on the public lands. The legislation aimed to eliminate indiscriminate settlement and grazing, stabilize the industry, and restore damaged lands. A Division of Grazing was created within the Department of the Interior. Grazing allotments and fees were established and leases issued to financially secure landowners who could take advantage of the opportunity. Only those with "base properties"—large private ranches near the public lands—were eligible to obtain leases. The Taylor Grazing Act maintained the large landowners in a dominant position, allowing small-scale graziers or newcomers to the industry no opportunity to obtain permits.

In 1946 the grazing lands were placed under the jurisdiction of the Bureau of Land Management (BLM), but little changed. Other than the elimination of roving bands of cattle herders across the West, little law enforcement or regulation ensued. The status quo had not changed, but violent outbursts between large operators and interlopers had been minimized by putting all the lands in the hands of lease holders. For some that was a situation not without merit, because the depression had pushed many homeless Americans onto the public lands where they were living as squatters, a situation untenable to the federal government and fraught with problems if the government were forced to remove settlers, or if anarchy developed.

The West is home to public lands ranching largely because that is where the public lands are. The federal government owns millions of acres of the West, and so do state and local agencies. Much of this land is remote, desert, or unsuited for cultivation or settlement. Grazing interests and environmental interests continue to collide over its use, but grazing on public lands has little support from other parts of the cattle industry. Low- or no-cost grazing only tips the balance against small operators and landowners who cannot obtain leases, making it a wedge driven into the "free market" of Western beef growers.

Western grazing, whether on private or public lands, supports a very small proportion of the total U.S. cattle market. The number of cattle on Western ranges pales in comparison to other parts of the nation—for instance Alabama, where pasturage supplies the same amount of beef and mutton as *all* the Western public lands.

Wisconsin has three times as many cows as Wyoming; Nebraska raises nearly seventeen times the number of cattle as Nevada. Most Western rangeland is too arid, too rocky, and too harsh in climate to provide good forage and conditions for cattle. It takes 1 acre per year of Iowa land to graze a cow but 230 acres of Nevada range land. An average comparison is 5 acres per year of Eastern grazing land to equal 185 acres of Western BLM and Forest Service leased land.

If Western public lands are less productive and the number of cattle grazing on them is negligible to the national food supply (less than 2 percent of the beef supply), why are critics so adamantly against cattle on public lands? Because if too many animals graze an area, the vegetation does not grow back as quickly nor as completely. Erosion from wind and rain ensues, precious topsoil washes away. Overgrazing, combined with an arid climate and drought, can turn an area into desert.

Cattle grazing can also produce more tree seedlings, something that both annoys and pleases scientists, depending upon their role within the U.S. Forest Service. Rangeland scientists want to see more grass and fewer trees, while forestry experts are delighted that cattle eat away the grass, allowing more seedlings to grow.[12] Wendee Holtcamp, in an article in *American Forests*, claims that "when cows trample and graze understory grasses and herbs, they remove competition for young saplings, allowing young trees to grow in unnaturally high densities."[13] Cattle have become an environmental football, tossed between opposing interests within the USDA.

CATTLE GRAZING has an enormous impact on plant life, just as the herds of millions of buffalo once did. One hundred fifty years ago there were between 40 and 75 million buffalo in North America, 7.5 million in the Western states, most of them on the plains of Montana, Wyoming, and Colorado. Herds of millions of animals stretched 100 miles or more in length. They ravished the plains as they passed through. Their wallowing in waterways left thousands of ponds in their wake, which turned to mud, then dust. Some old wallowing holes are still visible today. As the buffalo diminished, so did their use as a food source for plains Indians, who were eventually moved to reservations where they received beef allotments as subsistence, from cattle raised on the rangelands or trailed north from Texas. Today buffalo are raised on private ranches and are protected in national parks; about 150,000 bison range the West.

The bison at Yellowstone National Park present an interesting situation in which the borders between domestic and wild animals are continually fraught with conflict. Bison in the park are allowed to range freely, but when they encounter their bovine cousins, range cattle, they become a problem. An activist group called Buffalo Field Campaign maintains a close watch on bison that wander out of Yellowstone, herding them back inside the boundaries to protect them from Montana state agents who must destroy the animals if they are suspected of being infected with brucellosis. Brucellosis is a bacterial disease that the buffalo contracted from cattle; more than half the Yellowstone bison are positive for the disease, but only a few are carriers. Nevertheless brucellosis can be deadly to cattle and can sicken humans.

Montana officials claim they are between a rock and a hard place with Yellowstone's wandering bison. Unless they control them and the spread of brucellosis, other states will refuse to allow Montana beef in their markets. Volunteers at the Buffalo Field Campaign, a nonprofit operation with a budget of $10,000 a month, follow the bison in winter, shooing them back into the park boundaries if they wander. "Basically, you just throw your hands up and make a little noise and detour them back the way they came," a BFC patrol activist explains. To Montana's cattle raisers, the bison are symbols not of nature but of much-resented federal authority. The feds are blamed for the roaming bison. "Clearly this fight isn't just about disease eradication; it's about power and the control of public lands—an old and bitter issue in the American West," notes an article in *Mother Jones* magazine.[14] The issue is brucellosis but much more. It also concerns who manages wildlife and how the division between wild and domestic animals is worked out on both public and private lands in the West.

The Buffalo Field Campaign volunteers, made up of 250 people of many backgrounds and ages (though few are from Montana), maintain close ties to the bison. They interact with them in ways similar to cattle owners and their herds: "We spend hours with the buffalo, days and months on end; we name some of them. And having to watch them get thrown in a capture facility and get carted off [by Montana authorities] for some mindless reason, it's real depressing," one told a reporter. The U.S. Animal Health Association recommends a logical solution: eliminate brucellosis in the bison. That would mean rounding them up, killing animals that tested positive for the disease, and vaccinating the rest. It would end the standoff but create a culturally untenable situation: bison, emblematic of the American "wild," would essentially become "domestic," something few environmental activists can accept. Although it would create a healthy situation for the bison, their health is not the focus of concern. Their value as a symbol would be gone because dosing them with vaccines would "essentially destroy the wild nature of America's most cherished bison herd," according to *Mother Jones*.[15] Instead the federal government has purchased an adjoining ranch for $13 million to create a buffer zone that allows migrating buffalo to wander beyond the park's boundaries without creating problems for Montana's cattle raisers—who regard the federal government as "anti-cow."

Is there any common ground between environmental advocates and ranchers? Activists such as Lynn Jacobs propose citizen action to reform the way public lands are managed, and he is correct—there is an urgent need to ensure that the lands are maintained as carefully as possible. The nation wants to preserve nature, not see it destroyed by private interests, whether grazing, logging, or mining. Since the public has little authority to tell private landowners what to do, the solution many activist organizations have adopted is to become ranch owners too, by purchasing large tracts of land and then doing nothing with them. Allowing land to return to a "natural" state, grazed only by wild animals, is effective in some ways but is eeerily reminiscent of the Highland clearances in Scotland two centuries ago, when landowners pushed rural residents and small villages off the land, buying up larger and larger tracts in order to graze sheep. Eventually wool declined in value as the American South produced more cotton. Large tracts were then returned to the wild and enjoyed as game preserves for the elites.

In Africa such clearances are causing bloodshed and anarchy, as pastoral herding people are being pushed off their traditional lands to create government game preserves for tourists and large corporate-owned farms for investors. Five hundred thousand Wakamba in Kenya fight to maintain their grazing lands, but they, like other native tribes in Africa, are being crowded onto marginal lands where they end up overgrazing, turning the environment to desert, and ultimately starving. In 1999 the Kenyan government decided to cut private cattle holdings in half. It forced the frustrated Wakamba graziers to sell thousands of cattle at a loss. As Jomo Kenyatta writes in the *New Statesman*, "A European company has erected a factory for the canning of beef and other meat products on land adjoining the Athi River Station. It seems that, as a result, efforts are being made by the administration to ensure a steady supply of cattle for slaughter at that factory. . . . Pressure is being brought to bear on our tribe to dispose of our stock. . . . In a country which lives by cattle-farming, few things could undermine trust in the Government more fundamentally."[16] In Botswana, too, a similar effort is underway, where the government has enforced a system of privatized commercial beef ranches with fenced pastures. Fencing is the antithesis of pastoral grazing, and again, the conservation of overgrazed lands is the rationale. In Botswana, wildlife conservation is a prominent national policy: 17 percent of the country is set aside in national parks, and another 22 percent is in wildlife management areas.[17] Rural livestock owners are being displaced as the country moves along the same historical pattern: increased urbanization, with domestic livestock removed from public lands in favor of wildlife.

For the herders, however, it appears to be a power struggle between the haves and have-nots. Fencing, for example, is required in order to export beef to the European Union; it minimizes foot-and-mouth disease problems. J. S. Perkins, at the University of Botswana, points out that the situation is portrayed as one of overgrazing while in reality it is "being driven by socio-economic inequalities that are amongst the most pronounced in the world." Deepening rural poverty and urban migration are the issues, according to Perkins, not overgrazing—but it is a useful tool for pushing communal cattle operations out of business.[18]

Even traditional cattle raiding has been affected by Africa's growing urbanization. Cattle clans from Uganda and Kenya were battling it out in May 2000, crossing the borders to steal cattle as they historically have done. But this is a new millennium, so the raiders carried weapons purchased from Sudanese rebels. Cattle rustlers armed with AK-47 rifles and land mines attacked and raced home with two thousand head of cattle. They no longer keep the animals as wealth or use them as bridal dowries; the cattle are slaughtered and sold in urban areas.[19] Social dysfunction sets in as the cattle culture collapses. Cattle once were necessary as a dowry, which limited population growth. Without such a social constraint on reproduction, overpopulation overtakes resources. Cattle once were essential to diet and well-being, and provided a hedge against the future. Without their cattle and ancestral grazing lands, the people become angry refugees pushed to overcrowded urban areas.

The same dynamics are sometimes played out in the United States—or at least people worry that they might be. In small rural Western towns, fear surfaces when outsiders or the government threaten to quell what little economic activity exists, in most cases what revolves around agriculture or cattle. The interference usually stems from environmental concerns. Returning land to the "wild" forces local families and merchants to move away.

Rural residents are familiar with mean-spirited and illegal behavior in the guise of environmentalism. The book *Waste of the West* ends by urging anti-grazing activists to "take the Bull by the horns." It describes how anti-grazing activists "have begun sabotaging—*monkeywrenching*—the machinery and developments that enable the ranching establishment to ravage public land." Beneath a vivid illustration of a person wearing a ski cap and cutting a rancher's fence with wire-cutters by moonlight, a caption reads: "They cut fences; leave gates open; drive cattle onto neighboring allotments; decommission destructive ranching machinery; damage pumps, windmills, and stock tanks; dismantle and burn corrals, pens, and ramps; close ranching roads; leave stop-ranching messages on livestock road signs; dispose of salt blocks; remove traps and poisons; and generally do what they can to thwart the industry's ability to continue business-as-usual."[20] The book's litany of suggestions for anarchy against one's neighbor is nothing new, in fact "monkeywrenching" is rooted in the cattle culture's past. It is what clans did to each other in the cattle wars of northern England, Scotland, Ireland, and Africa. The fight is over cattle and land, and ultimately power.

Activism against cattle and their owners affects the animals as well as the participants. In *Tony and the Cows: A True Story from the Range Wars*, Will Baker writes about how monkeywrenching can ultimately be bad for one's soul. He explores the way eco-moralism has become an integral part of the environmental protection movement. As a freelance journalist sent to cover an Earth First! rendezvous in northern California for a New York magazine, Baker found disturbing events evolving out of the environmental movement's stance toward nature and domestic life. One of the Earth First! members from New Mexico, Tony Merten, was later suspected of killing several dozen cattle grazing on remote public lands. After questioning by local authorities, Merten committed suicide in his isolated cabin. Baker had communicated with Tony Merten after meeting him in California. "Tony thought wild lands ought to be reserved for wild creatures. Cattle, he remarked, were European in origin and did not belong on this continent," Baker writes.[21] This notion, that cattle are destructive alien creatures that wreak havoc on North American ecosystems, is peculiar to late-twentieth-century thinking. "Cattle are mentioned fifty-one times in the Book of Genesis," Baker argues. They have long been a symbol and source of sustenance and wealth, and have been worshiped by non-Christian religions for thousands of years.

But, as Baker points out, a nostalgia for a poignant, pastoral dreamworld has enveloped Americans, and their yearning for escape from urban toil has been "transformed into a new set of fantasies, lifestyles, and scenarios: outdoor adventure, country estates, survival experiences."[22] This nostalgic vision of "nature" has created a surge of interest in simple, primitive ways. Perhaps not exactly in *living* that way (no one wants to go into the wild without their Powerbook, synthetic boots, or SUV) but in extolling the *virtues* of living that way.

But there is an emotional and psychological aspect of "ecowar" that has not been addressed by the ideologists of the environment, whose ideas affect people like Tony Merten, who could do nothing to defend his principles but kill cattle and ultimately himself. They feel doomed and helpless because they are humans who have sworn off a domestic nature, and therefore off of humanity. Baker looks at how Ted Kaczynski's crude cabin replicated Thoreau's model for living a life removed from the clutter and banality of domestic life. Is a self-made shanty a mark of a higher, enlightened life? Is solitude and the refutation of progress the mark of genius? A bevy of writers have followed Thoreau's example. They write on lonely backroads or slickrock canyons, a "horde of recent scribblers" who set out alone to confront nature with a journal. Commendable, of course, but the same writers are quick to criticize others who write about cowboys and cattle in the domesticated West.

So few Americans would give up their house, car, pool, PC, and TV for a wilderness shack and a bean field. "Hence the wilderness functions more and more in a dual role," Baker points out, "as research station and religious theme park."[23] Nature has become a shrine, "carefully preserved and museum-like, sometimes with ticket booths and convenience stores at the gate."[24] What has happened in the West is similar to Buffalo Bill's Wild West Show of the nineteenth century—a synthesized version of the "wild" maintained for entertainment: survival camps, river guides, game ranches, bicycle and hiking trails. They allow one to reenact the natural world that now exists only in mythology. Self-reliant loners, trying to live outside the "herd," make romantic fictional characters, but we are a community of people, and our domestic livestock is part of that community.

The fight between ranchers and environmentalists will continue because it envelops a contrast in moral vision. The national soul is torn between the domestic and the wild. The public, whose opinion is the prize, values wilderness as a symbol and hamburger as a meal. So "if we are to save ourselves and this little atom we inhabit, we must do so together," Will Baker asserts. Wearing T-shirts that proclaim "My heroes have always killed cowboys," gets attention, but it may lead farther than society wants to go.[25]

BARNEY NELSON teaches English and literature at Sul Ross State College in Alpine, Texas, and describes herself as "a genuine crusty old woman from the rural culture."[26] Crusty or not, she embraces the rural West in a way few female writers have been able to do. Nelson examines the way Western literature and myth have created a dichotomy between the wild and the domestic, which she points out has preempted women from any place in today's wilderness ideologies. She notes that Thoreau's 1862 essay, "Walking," has been reshaped by modern scholars to omit all of Thoreau's references to domestic livestock and has been used to promote a male-dominated "natural" world. The omission of his references to domestic animals, farming, and pastures has reshaped his work into what Nelson calls "wilderness preservation propaganda." Domestic livestock and farming have been equated with the feminine while wilderness has been seen as a man's domain. The ideology of today's environmentalism revolves around a "wild" in nature that cannot be tainted by the domestic: the female, the family, the livestock. History and literature, says Nelson, are filled with what is largely ignored, what she calls "the lost domestic."[27]

In "Walking," Nelson writes, Thoreau "reminds us that the cow was not born domestic, but is in fact descended, like all people, from the savage." He wrote about observing cattle and how closely they resembled wild animals; they had not lost that spark, that "seed of instinct" and their "wild habits and vigor." But, as Nelson points out, Thoreau's writing has today been refined, polished, and made to correspond to today's vision of the wild, as a way to gain support for preservation politics.[28] This struggle between ideas of the wild and the domestic has created a cultural divide between urban and rural people worldwide. The attitude that wild animals are easier on the environment than domestic ones extends even to the third world. There the idea that wild animals consume fewer resources than domestic ones can devalue third world populations by assuming they can (and should) exist on a fraction of the resources of those living in more developed countries. In the American West, environmentalists sincerely believe they are helping the family farmer, but they play into the hands of agribusiness by demanding more regulations, which inevitably force smaller players out of the industry. Urban writers, according to Nelson, have denigrated the people, animals, and lifestyles of the rural, whose people have responded by "opposing the government, the universities, public education in general, and 'environmentalists,' a vague term applied to anyone who disagrees with them."[29]

American mythology is filled with images of strong male leaders with survival skills who seek the wild, free from boring domestic restraints such as rearing children. Wild bachelorhood has been idealized in the films and novels of the West, in the character of mountain men, cowboys, and frontiersmen—all handsome, self-reliant, and single. The males of Zane Grey (a New York dentist), Frederic Remington (a New York painter), and Owen Wister (a summer visitor to the West for his health), along with John Muir and Jack London's men, are virile and indestructible. Animals, when portrayed in their literature, are like the male stereotype: the lone wolf, stallion, or bull. Western men are seldom portrayed helping women get food or caring for the young. The image of cattle, of a herd led by older, aggressive females, with group energy spent protecting the young, does not fit the mythological "wild" West.

CATTLE have come on a long journey with us, from pastoral times to settled agriculture, from the New World to post-industrialism. From the days of dried jerky, to canned milk, to whey protein concentrates, they have nourished us. As we enter an age of biotechnology with them (one cow has already been cloned in Japan, others will follow), it is well to remember that we have a bond with cattle, a responsibility. Stephen Budiansky reminds us we have "a duty to animals who have cast their lot with ours and who by their evolutionary heritage are by nature dependent upon us."[30]

Any rancher will admit that cattle are not really "tame." They continue to behave as cattle naturally do. Some have better temperaments than others, but even a bottle-fed calf can grow into a bull that is deadly. They do not become pets, nor do they show affection for humans as horses or dogs do. More like cats, they have stuck with us because we feed them. Shaping their environment with technology and government, and changing our lives to fit theirs, makes it possible to enjoy their benefits. Obtaining hay and feed, building pens and barns, and learning how to preserve milk and its products has changed us far more than it has changed the bovines. One could argue that *they* domesticated us.

Acknowledgments

THE IDEA FOR THE BOOK originated in a rural history class taught by David Coon at Washington State University. Attending class at Washington State was an inspiration to me, not only because the libraries held so much valuable information about cattle, but because each time I drove the seventy-five miles to Pullman I passed several cattle ranches. I began to watch eagerly for a glimpse of the new calves and noticed how a herd composed itself during grazing. The new mother cows would graze together, the babies all securely under the watchful eye of a few older cows, in what passes for cow day-care. Bulls, in expansive pastures, hung out together too, in a boys' club of sorts. If it was afternoon, all were lying in the shade, chewing cud. The slaughter truck made its appearance now and then. But the tempo of the seasons and the continuity of herd life struck me every time I made the trek to class.

I found that there is an entire cattle culture in the West that runs "below the radar" for most of us. When I made research calls one spring, Page Lambert informed me that everyone on the plains was in the middle of calving. Yet, like many others, she took time to direct me to useful contacts and information. There are many, many stories out there—too many for this book. Nevertheless I thank the following for ideas and comments they shared which shaped my thinking as I approached the subject: Judy Dammel, Linda Hasselstrom, Nancy Curtis, Gaydell Collier, Tallie Thompson, Jerri Stone, Jerrie Hurd, Hugh Williams, Kelley Pounds,

Corinne Brown, Paige Palmer Nelson, Persia Woolley, Gail Fiorini-Jenner, and Rachel Klippenstein. The response from members of Women Writing the West was invaluable.

The following people also willingly gave important advice and direction, and I appreciate their help very much: David Coon, Michael Egan, Paul Hirt, Susan Armitage, Michael Green, Donald Nelson, John R. Gorham, Fred Lauritsen, Claude Nichols, Jerri Bartholomew, Donald A. Henderson, Frank Fenner, Marilyn Carlsen, David Smestad, Judith Keeling, Trevor Bond, June Sawyers, and, of course, my publisher, Ivan Dee.

I owe many thanks to my family for their support and willingness to listen and critique my ideas: John, Ed, Molly, and my parents Ed and Juanita. Terry was a very supportive spouse; he willingly fenced our acreage, baled the hay, and has been doing the heavy lifting for our feeder steer research project. And the three steers have been happy to go along with it all—so far.

Because there is no other place to say this, let me mention too the Heifer Project, a commendable charitable idea that helped me realize how owning livestock empowers impoverished people, particularly women and their families.

L. W. C.

Cheney, Washington
June 2001

Notes

Chapter 1. Cows on the Ceiling

1. Paul G. Bahn, *The Cambridge Illustrated History of Prehistoric Art* (New York: Cambridge University Press, 1998), 58.

2. Bahn, *Illustrated History*, 154.

3. Pedro A. Saura Ramos, *The Cave at Altamira* (New York: Harry M. Abrams, 1998), 27.

4. Saura Ramos, *Cave at Altamira*, 22.

5. Saura Ramos, *Cave at Altamira*, 24.

6. Bahn, *Illustrated History*, 126.

7. N. K. Sandars, *Prehistoric Art in Europe* (New York: Penguin Books, 1985), 101.

8. Saura Ramos, *Cave at Altamira*, 79.

9. Bahn, *Illustrated History*, 173.

10. Bahn, *Illustrated History*, 67.

11. Bahn, *Illustrated History*, 221.

12. Ibid.

13. L. G. Freeman, J. Gonzalcz Echcgaray, F. Gernaldo de Quirós, and J. Ogden, *Altamira Revisited, and Other Essays on Early Art* (Chicago: Institute for Prehistoric Investigations, 1987), 71.

14. Freeman, et al., *Altamira Revisited*, 92.

15. Freeman, et al., *Altamira Revisited*, 77.

16. Freeman, et al., *Altamira Revisited*, 239.

17. Sandars, *Prehistoric Art in Europe*, 115.

18. Sandars, *Prehistoric Art in Europe*, 142.

19. Sandars, *Prehistoric Art in Europe*, 171.

Chapter 2. The Domestication of Cattle

1. Frances Galton, "Gregariousness in Cattle and Men," *Macmillan's Magazine* 23 (1871), 353.

2. Carl O. Sauer, *Agricultural Origins and Dispersals: The Domestication of Ani-*

mals and Foodstuffs (Cambridge: Massachusetts Institute of Technology Press, 1969), 86.

3. Sauer, *Agricultural Origins*, 89.

4. Sauer, *Agricultural Origins*, 97.

5. Frederick E. Zeuner, *A History of Domesticated Animals* (New York: Harper and Row, 1963), 35.

6. Jared Diamond, *Guns, Germs, and Steel* (New York: W. W. Norton, 1997), 166.

7. S. Boyd Eaton, Marjorie Shostak, and Melvin Konner, *The Paleolithic Prescription: A Program of Diet and Exercise and a Design for Living* (New York: Harper and Row, 1988), 72.

8. Sauer, *Agricultural Origins*, 42.

9. Zeuner, *History*, 220.

10. Zeuner, *History*, 225.

11. Zeuner, *History*, 228.

12. Zeuner, *History*, 244.

13. Stephen Budiansky, *The Covenant of the Wild: Why Animals Chose Domestication* (New York: William Morrow, 1992), 15.

14. Budiansky, *Covenant*, 24.

15. Budiansky, *Covenant*, 62.

16. Edward O. Wilson, *Sociobiology* (Cambridge, Mass.: Harvard University Press, 1980), 38.

17. Galton, "Gregariousness in Cattle and Men," 354.

18. Ibid.

19. Ibid.

20. Ibid.

21. Galton, "Gregariousness in Cattle and Men," 356.

Chapter 3. Cattle and the Clan

1. Budiansky, *Covenant*, 142. Chagnon's research methods have been questioned in part because his findings did not agree with our image of how hunting and gathering cultures *should* behave.

2. Donald Worster, "Nature and the Disorder of History," *Environmental History Review*, Summer 1994, 12.

3. Worster, "Nature and Disorder," 13.

4. Michael Richards, "First Farmers with No Taste for Grain," *British Archaeology* 12 (March 1996), 2.

5. Julian Morgenstern, *The Book of Genesis: A Jewish Interpretation* (New York: Schocken Books, 1965).

6. Bruce Lincoln, *Priests, Warriors, and Cattle* (Berkeley: University of California Press, 1981), 7.

7. Lincoln, *Priests, Warriors, and Cattle*, 16.

8. Lincoln, *Priests, Warriors, and Cattle*, 10.

9. Craig B. Stanford, *The Hunting Apes: Meat Eating and the Origins of Human Behavior* (Princeton, N.J.: Princeton University Press, 1999), 5.

10. Stanford, *Hunting Apes*, 41.

11. Stanford, *Hunting Apes*, 90.

12. Robert K. Logan, *The Alphabet Effect: The Impact of the Phonetic Alphabet on the Development of Western Civilization* (New York: William Morrow, 1986), 22.

13. Logan, *Alphabet Effect*, 37.

14. Albertine Gaur, *A History of Writing* (New York: Cross River Press, 1992), 127.

15. Gaur, *History of Writing*, 45.

Chapter 4. Woman's Best Friend

1. Dirk van Loon, *The Family Cow* (Pownal, Vt.: Storey Books, 1976), 239.

2. Barbara G. Walker, *The Woman's Encyclopedia of Myths and Secrets* (San Francisco: HarperCollins Publishers, 1983), 181.

3. Stanford, *Hunting Apes*.

4. Lincoln, *Priests, Warriors, and Cattle*, 177.

5. Walter L. Brenneman, Jr., "Serpents, Cows, and Ladies: Contrasting Symbolism in Irish and Indo-European Cattle-Raiding Myths," *History of Religions* 28:4 (May 1989), 340.

6. Brenneman, "Serpents, Cows, and Ladies," 344.

7. Brenneman, "Serpents, Cows, and Ladies," 348.

8. Duncan Norton-Taylor, *The Celts*, The Emergence of Man Series (New York: Time-Life Books, 1974), 85.

9. Norton-Taylor, *Celts*, 48.

10. Norton-Taylor, *Celts*, 73.

11. Thomas Cahill, *How the Irish Saved Civilization: The Untold Story of Ireland's Heroic Role from the Fall of Rome to the Rise of Medieval Europe* (New York: Doubleday, 1995), 81.

12. Cahill, *How the Irish Saved Civilization*, 101.

13. Cahill, *How the Irish Saved Civilization*, 89.

14. *Celtic Myth*, Heroes of the Dawn Series (London: Duncan Baird Publishers, 1996), 37.

15. *Celtic Myth*, 42.

16. Cahill, *How the Irish Saved Civilization*, 77, 127.

17. Thomas Kinsella, *The Tain* (New York: Oxford University Press, 1970), xiii.

18. T. W. Rolleston, *Myths and Legends of the Celtic Race* (London: George G. Harrap, 1911), 202.

19. Rolleston, *Myths and Legends*, 203.

20. Cahill, *How the Irish Saved Civilization*, 175.

21. William W. Fitzhugh and Elisabeth I. Ward, eds., *Vikings: The North Atlantic Saga* (Washington, D.C.: Smithsonian Institution Press, 2000), 71.

22. Catherine Manton, *Fed Up: Women and Food in America* (Westport, Conn.: Bergin and Garvey, 1999), 32.

23. Joan M. Jensen, *With These Hands: Women Working on the Land* (Old Westbury, N.Y.: Feminist Press, 1981), 107.

24. Jensen, *With These Hands*, 281.

25. Jensen, *With These Hands*, 112.

26. Jensen, *With These Hands*, 199–202.

27. J. H. M. Pinkerton, *Advances in Oxytocin Research* (New York: Pergamon Press, 1965), 147.

28. Linda Hasselstrom, Gaydell Collier, and Nancy Curtis, eds., *Leaning into the Wind: Women Write from the Heart of the West* (Boston: Houghton Mifflin Co., 1997), xviii.

29. Hasselstrom, et al., *Leaning into the Wind*, xv.

30. Hasselstron, et al., *Leaning into the Wind*, 133–134.

Chapter 5. Cattle Culture Comes to the Americas

1. Alexis de Tocqueville, *Democracy in America* (New York: Alfred A. Knopf, 1976).

2. This is based on the spatial land-rent model developed by German scholar Heinrich von Thunen, in Terry G. Jordan, *North American Cattle-Ranching Frontiers; Origins, Diffusion, and Differentiation* (Albuquerque: University of New Mexico Press, 1993), 11.

3. Jordan, *North American Cattle-Ranching Frontiers*, 14.

4. Jordan, *North American Cattle-Ranching Frontiers*, 17.

5. Terry G. Jordan, *Trails to Texas: Southern Roots of Western Cattle Ranching* (Lincoln: University of Nebraska Press, 1981), 156.

6. Jordan, *Trails to Texas*, 14.

7. David L. Coon, *The Development of Market Agriculture in South Carolina*, 1670–1785 (New York: Garland Publishing, 1989).

8. Jordan, *Trails to Texas*, 34.

9. Jordan, *Trails to Texas*, 33.

10. William Brohaugh, *English Through the Ages* (Cincinnati: Writers Digest Books, 1998), 182.

11. John McPhee, *The Crofter and the Laird* (New York: Farrar, Straus and Giroux, 1970), 7.

12. Ibid.

13. McPhee, *Crofter*, 137.

14. Peter Wilson Coldham, *Bonded Passengers to America: Western Circuit, 1664–1775* (Baltimore: Geneological Publishing Co., 1983), V, xiv, 31.

15. Jacqueline A. Rinn, "Scots in Bondage: Forgotten Contributors to Colonial Society," *History Today* 30 (1980), 18.

16. Rinn, "Scots in Bondage," 20.

17. Rinn, "Scots in Bondage," 21.

18. Forrest McDonald and Grady McWhiney, "The Celtic South," *History Today* 30 (1980), 11.

19. David Hackett Fischer, *Albion's Seed: Four British Folkways in America* (New York: Oxford University Press, 1989), 25.

20. McDonald and McWhiney, "Celtic South," 11.

21. Forrest McDonald and Grady McWhiney, "Celtic Origins of Southern Herding Practices," *Journal of Southern History* 60:2 (1985), 189.

22. Ibid.

23. Daniel Akst, "The Forgotten Plague," *American Heritage* 51:8 (2000), 72.

24. John Solomon Otto, "The Migration of the Southern Plain Folk: An Interdisciplinary Synthesis," *Journal of Southern History* 60:2 (1985), 189.

25. McDonald and McWhiney, "Celtic Origins of Southern Herding Practices," 168.

26. McDonald and McWhiney, "Celtic Origins of Southern Herding Practices," 173.

27. Otto, "Migration," 189, 183.

28. Otto, "Migration," 199.

29. Sarah Richardson, "Vanished Vikings," *Discover*, March 2000, 69.

30. Jim Patrico, "Cattle Come to America: Five Hundred Years of Beef in the New World," *Beef Today*, March 1994, 63.

31. Patrico, "Cattle," 64.

32. Ibid.

33. Fischer, *Albion's Seed*, 626.

34. Fischer, *Albion's Seed*, 639.

35. Fischer, *Albion's Seed*, 659.

36. Fischer, *Albion's Seed*, 662.

Chapter 6. From Trails to Tracks

1. George Jackson, *Sixty Years in Texas* (Dallas, 1908), 11.

2. Cecil Kirk Hutson, "Texas Fever in Kansas, 1866–1930," *Agricultural History* 68:1 (1994), 77.

3. J. Frank Dobie, *The Longhorns* (New York: Little, Brown, 1941), 5.

4. Dobie, *Longhorns*, 11.

5. Dobie, *Longhorns*, 18.

6. Dobie, *Longhorns*, 23.

7. Joseph G. McCoy, *Historic Sketches of the Cattle Trade of the West and Southwest* (Washington, D.C.: Rare Book Shop, reprint, 1932), 20.

8. McCoy, *Sketches*, 38.

9. McCoy, *Sketches*, 23.

10. McCoy, *Sketches*, 53.

11. Ibid.

12. McCoy, *Sketches*, 56.

13. McCoy, *Sketches*, 57.

14. McCoy, *Sketches*, 64.

15. McCoy, *Sketches*, 65.

16. McCoy, *Sketches*, 66.

17. McCoy, *Sketches*, 73.

18. Ibid.

19. McCoy, *Sketches*, 77.

20. McCoy, *Sketches*, 86.

21. McCoy, *Sketches*, 95.

22. McCoy, *Sketches*, 99.

23. McCoy, *Sketches*, 115, 116.
24. McCoy, *Sketches*, 153.
25. McCoy, *Sketches*, 159.
26. McCoy, *Sketches*, 158.
27. McCoy, *Sketches*, 161.
28. McCoy, *Sketches*, 162.
29. McCoy, *Sketches*, 25.
30. Hutson, "Texas Fever in Kansas," 80.
31. Claire Strom, "Texas Fever and the Dispossession of the Southern Yeoman Farmer," *Journal of Southern History* 66:1 (2000), 53.
32. Strom, "Texas Fever," 56.
33. Strom, "Texas Fever," 56.
34. Strom, "Texas Fever," 61.
35. Strom, "Texas Fever," 68.
36. Strom, "Texas Fever," 68.
37. Strom, "Texas Fever," 74.
38. Strom, "Texas Fever," 64, 73.

Chapter 7. The Devil's Rope

1. Walter Prescott Webb, *The Great Plains* (Waltham, Mass.: Blaisdell Publishing Co., 1931), 239.
2. Webb, *Great Plains*, 234.
3. Webb, *Great Plains*, 287.
4. Henry D. McCallum and Frances T. McCallum, *The Wire That Fenced the West* (Norman: University of Oklahoma Press, 1965), 23.
5. Webb, *Great Plains*, 292.
6. McCallum, *Wire That Fenced the West*, 25.
7. Webb, *Great Plains*, 292.
8. McCallum, *Wire That Fenced the West*, 27.
9. McCallum, *Wire That Fenced the West*, 31.
10. McCallum, *Wire That Fenced the West*, 107.
11. McCallum, *Wire That Fenced the West*, 66.
12. David Dary, *Entrepreneurs of the Old West* (Lincoln: University of Nebraska Press, 1986), 247.
13. McCallum, *Wire That Fenced the West*, 133.
14. Oscar H. Flagg, *A Review of the Cattle Business in Johnson County, Wyoming, Since 1892 and the Causes That Led to the Recent Invasion* (New York: Arno Press, 1969), 1.
15. Flagg, *Review*, 8.
16. Flagg, *Review*, 5.
17. Flagg, *Review*, 19.
18. Flagg, *Review*, 21.
19. Ezra Bowen, ed., *The Cowboys*, The Old West Series (Alexandria, Va.: Time-Life Books, 1978), 61.

20. Paul Johnson, *A History of the American People* (New York: HarperCollins, 1997), 516.

21. "Tribute to Barbed Wire," Internet, at www.barbwiremuseum.com/BW-Monument.htm

Chapter 8. The Industrialization of Meat

1. James R. Barrett, *Work and Community in the Jungle: Chicago's Packinghouse Workers, 1894–1922* (Chicago: University of Chicago Press, 1990), 20.

2. William Cronon, *Nature's Metropolis: Chicago and the Great West* (New York: W. W. Norton, 1991), 207.

3. Cronon, *Nature's Metropolis*, 211.

4. Cronon, *Nature's Metropolis*, 233.

5. Cronon, *Nature's Metropolis*, 234.

6. Cronon, *Nature's Metropolis*, 235.

7. Cronon, *Nature's Metropolis*, 241.

8. Cronon, *Nature's Metropolis*, 244.

9. T. K. Derry and Trevor I. Williams, *A Short History of Technology from the Earliest Times to A.D. 1900* (New York: Dover Publishing Co., 1993), 699.

10. Samuel Plimsoll, *Cattle Ships* (London: n.p., 1890), 9.

11. Plimsoll, *Cattle Ships*, 87.

12. Cronon, *Nature's Metropolis*, 250.

13. Cronon, *Nature's Metropolis*, 253.

14. Albert Sonnenfeld, ed., *Food: A Culinary History from Antiquity to the Present* (New York: Columbia University Press, 1999), 488.

15. James Harvey Young, *Pure Food: Securing the Federal Food and Drugs Act of 1906* (Princeton, N.J.: Princeton University Press, 1989), 108.

16. Young, *Pure Food*, 111.

17. Margaret Leech, *In the Days of McKinley* (New York: Harper, 1959), 300.

18. Young, *Pure Food*, 138.

19. Young, *Pure Food*, 136.

20. Graham A. Cosmas, *An Army for Empire: The United States Army in the Spanish-American War* (Columbia: University of Missouri Press, 1971), 287.

21. Cosmas, *An Army for Empire*, 289.

22. Young, *Pure Food*, 139.

23. Sean Patrick Adams, "Hardtack, Canned Beef, and Imperial Misery: Rae Weaver's Journal of the Spanish-American War," *Wisconsin Magazine of History* (Summer 1998), 244.

24. Cosmas, *An Army for Empire*, 294.

25. *New York Times*, Feb. 28, 2001, B-12.

Chapter 9. Margarine: The Plastic Food

1. *The Kitchen Encyclopedia* (Chicago: Swift and Co., 1911), 2.

2. Maguelonne Toussaint-Samat, *History of Food*, trans. Anthea Bell (New York: Barnes and Noble Books, 1998), 122.

3. Toussaint-Samat, *History of Food*, 123.

4. Young, *Pure Food*, 72.

5. Mark Twain, *Life on the Mississippi* (New York: Book-of-the-Month Club, 1992, reprint of the 1883 ed.), 328–329.

6. Mitchell Okun, *Fair Play in the Marketplace: The First Battle for Pure Food and Drugs* (DeKalb, Ill.: Northern Illinois University Press, 1986), 253.

7. Young, *Pure Food*, 80.

8. Young, *Pure Food*, 263.

9. Young, *Pure Food*, 82.

10. Young, *Pure Food*, 82, 84, 86, 94.

11. Ruth Dupré, "If It's Yellow, It Must Be Butter: Margarine Regulation in North America Since 1886," *Journal of Economic History* 59:2 (1999), 353.

12. Dupré, "If It's Yellow," 359.

13. Dupré, "If It's Yellow," 360.

14. S. F. Riepma, *The Story of Margarine* (Washington, D.C.: Public Affairs Press, 1970), 17.

15. Wayne D. Rasmussen, ed., *Agriculture in the United States: A Documentary History*, 4 vols. (New York: Random House, 1975), 3621.

16. Riepma, *The Story of Margarine*, 6.

17. Dupré, "If It's Yellow," 370.

18. Erika von Mutius, et al., "Increasing Prevalance of Hay Fever and Atopy Among Children in Leipzig, East Germany," *The Lancet* 351 (1998), 865.

19. Ingo Witte, "Prevalence of Hay Fever and Consumption of Margarine in East Germany," *The Lancet* 352 (July 25, 1998), 318.

20. Jeffrey Kluger, "Margarine Misgivings," *Time* (July 5, 1999), 63.

21. Dawn Lyons-Johnson, "A Better Spread for Your Bread" (Agricultural Research Service, USDA, 1998).

22. Internet, www.margarine.org.

23. S. Delamarre and C. A. Batt, "The Microbiology and Historical Safety of Margarine," *Food Microbiology* 16:4 (1999), 327.

24. Rasmussen, *Agriculture in the United States*, 1689.

25. USDA Food and Agricultural Service, based on 1.97 kg. per capita.

Chapter 10. Free Speech and Hamburgers

1. "Oprah Says She Felt Redeemed After Victory over Texas Cattlemen," *Jet* (March 23, 1998).

2. Judith C. Juskevich and C. Greg Guyer, "Bovine Growth Hormone: Human Food Safety Evaluation," *Science* 249: 875–884.

3. Kelly Morris, "Bovine Somatotropin—Who's Crying Over Spilt Milk?" *The Lancet* 353:306.

4. Morris, "Bovine Somatotropin," 306.

5. F. Van Everbroeck, P. Pals, and P. Cras, "Safety Measures for Handling Laboratory Specimen and Patients with Creutzfeldt-Jakob Disease," *Ned Tijdschr Geneeskd* 143:29 (July 1999), 1511. (This information is available in English on the

National Library of Medicine database, via Medline, http://www.healthgate.com/ccgi-bin/q=format,cgi?f=G&d=fmb97&m=1059052&ui=99372220.)

6. Tamara Skilton, Washington State Funeral Director's Association, Bellevue, Wash. Phone conversation November 9,1999. Ms. Skilton noted that OSHA compliance regulations are the only regulations the industry falls under. Due to the lack of knowledge about CJD and its transmission, funeral homes in the state of Washington, and much of the nation, do not perform embalming services on bodies of those known to have been infected with CJD. Outside of hospitals, only two facilities in the nation have adequate protection for technicians to embalm CJD victims. The biggest concern is the number of Alzheimer's deaths that may be unrecognized cases of CJD.

7. Robert Berkow, ed., *The Merck Manual of Medical Information* (Whitehouse Station, N.J.: Merck Research Laboratories, 1997), 922.

8. J. W. Almond, "Bovine Spongiform Encephalopathy and New Variant Creutzfeldt-Jakob Disease," *British Medical Bulletin* 54 (1998), 750.

9. "Bovine Spongiform Encephalopathy (BSE) in the United Kingdom and Creutzfeldt-Jakob Disease (CJD) in the United States," Center for Disease Control, Internet available at http://www.cdc.gov/ncidod/diseases/cjd/qa96bse.htm.

10. Robert B. Petersen, "Antemortem Diagnosis of Variant Creutzfeldt-Jakob Disease," *The Lancet* 353 (Jan. 16, 1999), 164.

11. Berkow, *Merck Manual*, 922.

12. Howard F. Lyman and Glen Merzer, *Mad Cowboy: Plain Truth from the Cattle Rancher Who Won't Eat Meat* (New York: Scribner's, 1998), 14.

13. Ibid.

14. Sam Howe Verhovek, "A Gain for Winfrey in Beef-Producer's Suit," *New York Times*, Feb. 18, 1998.

15. The American Feed Industry Association is a lobbying group for manufacturers and distributors of animal feed additives such as microbials and chemicals. Another trade organization, the National Renderer's Association, also works to promote interests of the feed industry. Available from: http://www.rendermagazine.com/index.html.

16. "Farmer's Right to File Lawsuits Grows, Raising Debate on Safety of Food," *New York Times*, June 1, 1999, A1:C11.

17. Ibid.

18. Ibid; "Hot Type," *Chronicle of Higher Education*, June 30, 2000, A22.

19. "The Plot Against Alar: An ACSH Interview with Robert Bidinotto, Author of 'The Great Apple Scare' in *Reader's Digest*, Oct. 1990," American Council for Science and Health; http://www.acsh.org/publications/priorities/0301/alar.html.

20. Kerry E. Rodgers, "Multiple Meanings of Alar After the Scare: Implications for Closure," *Science, Technology, and Human Values* 21:2 (Spring 1996), 193.

21. Rodgers, "Multiple Meanings of Alar," 181.

22. Rodgers, "Multiple Meanings of Alar," 184.

23. Elliott Negin, "The 'Alar Scare' Was for Real, and So Is That 'Veggie Hate-Crime' Movement," *Columbia Journalism Review* (Sept./Oct. 1996), 15.

24. Jamison S. Prime, "Fruitfully Correct: Alar Incident Spurs States to Nip Critics in the Bud," *Quill* (Jan./Feb. 1995), 38.

25. Walter M. Brasch, *Forerunners of Revolution: Muckrakers and the American Social Conscience* (Lanham, Md.: University Press of America, 1990), 5.

26. Brasch, *Forerunners of Revolution*, 81.

27. Charles Edward Russell, *The Greatest Trust in the World* (New York: Ridgway-Thayer Co., 1905; reprint, New York: Arno Press, 1975).

28. Russell, *The Greatest Trust*, 213.

29. Ibid.

30. Russell, *The Greatest Trust*, 246.

31. Brasch, *Forerunners of Revolution*, 87.

32. Brasch, *Forerunners of Revolution*, 88.

33. Brasch, *Forerunners of Revolution*, 98.

34. Brasch, *Forerunners of Revolution*, 123.

35. Brasch, *Forerunners of Revolution*, 124.

36. Alexis de Tocqueville, *Democracy in America*, 1832. Available at http://xroads.virginia.edu/~HYPER/DEETOC/ch2_06.htm. Russell, *The Greatest Trust in the World*, 227.

37. Young, *Pure Food*, 282.

38. Bernard A. Weisberger, "Beef, Pork, and History," *American Heritage* (July/Aug. 1989), 22.

39. "Monsanto Statement Regarding the Oct. 26, 1999, Proposal on BST by the European Commission," Monsanto, October 27, 1999, ref. No. 2330. Available from Monsanto website: http://www.biotechknowledge.com/showlib_biotech.php3?2330.

40. Young, *Pure Food*, 291.

Chapter 11. Breeding Back the Aurochs

1. Heinz Heck, "The Breeding Back of the Aurochs," *Oryx* 1:3 (1951), 120.

2. Eugene M. Ensminger, *Beef Cattle Science*. Animal Agriculture Series (Danville, Ill.: Interstate, 1960), 6.

3. Derry and Williams, *Short History of Technology*, 70.

4. H. Cecil Pawson, *Robert Bakewell: Pioneer Livestock Breeder* (London: Crosby Lockwood and Son, 1957), 7.

5. Pawson, *Robert Bakewell*, 5.

6. Pawson, *Robert Bakewell*, 7.

7. Pawson, *Robert Bakewell*, 10.

8. Pawson, *Robert Bakewell*, 11.

9. Ibid.

10. Pawson, *Robert Bakewell*, 12.

11. Pawson, *Robert Bakewell*, 29.

12. Pawson, *Robert Bakewell*, 50.

13. Pawson, *Robert Bakewell*, 29.

14. Pawson, *Robert Bakewell*, 40.

15. Ibid.

16. Pawson, *Robert Bakewell*, 41.

17. Pawson, *Robert Bakewell*, 50.

18. Pawson, *Robert Bakewell*, 53.
19. Ensminger, *Beef Cattle Science*, 7.
20. Pawson, *Robert Bakewell*, 69.
21. Derry and Williams, *Short History of Technology*, 677.
22. Van Loon, *Family Cow*, 22.
23. Roger J. Wood, "Robert Bakewell: Pioneer Animal Breeder and His Influence on Charles Darwin," *Casopis Moravskeho Musea* 58:8 (1973), 236.
24. Charles Darwin, 1844, cited in Wood, "Robert Bakewell," 239.
25. Wood, "Robert Bakewell," 237.
26. Charles Darwin, 1877, cited in Wood, "Robert Bakewell," 239.
27. Wood "Robert Bakewell," 240.
28. S. J. Holmes, *Human Genetics and Its Social Import* (New York: McGraw-Hill, 1936), 57.
29. Thurman B. Rice, *Racial Hygiene: A Practical Discussion of Eugenics and Race Culture* (New York: Macmillan, 1929), xi.
30. Rice, *Racial Hygiene*, 94.
31. Rice, *Racial Hygiene*, 107.
32. Rice, *Racial Hygiene*, 159.
33. Ibid.
34. T. Swann Harding, "Eugenics for Cows But Not for Humans," *Scientific American* (January 1932), 26.
35. Ibid.
36. Harding, "Eugenics for Cows," 27.
37. Holmes, *Human Genetics and Its Social Import*, 359.
38. Holmes, *Human Genetics and Its Social Import*, 361.
39. Holmes, *Human Genetics and Its Social Import*, 147.
40. Heck, "Breeding Back of the Aurochs," 119.
41. Zeuner, *History of Domesticated Animals*, 204.
42. Heck, "Breeding Back of the Aurochs," 120.
43. Ibid.
44. Zeuner, *History of Domesticated Animals*, 206.
45. Kenneth M. Ludmerer, *Genetics and American Society: A Historical Approach* (Baltimore: Johns Hopkins University Press, 1972), 73.
46. Ludmerer, *Genetics and American Society*, 63.
47. Ludmerer, *Genetics and American Society*, 83.
48. Ludmerer, *Genetics and American Society*, 84.
49. Ludmerer, *Genetics and American Society*, 87.
50. Sheila Faith Weiss, "The Race Hygiene Movement in Germany, 1904–1945," in *The Wellborn Science: Eugenics in Germany, France, Brazil, and Russia*, ed. Mark Adams (New York: Oxford University Press, 1990), 38.
51. Weiss, "Race Hygiene Movement," 10.
52. Ludmerer, *Genetics and American Society*, 117.
53. Adams, *Wellborn Science*, 44.
54. Adams, *Wellborn Science*, 179.
55. Adams, *Wellborn Science*, 181.
56. Adams, *Wellborn Science*, 197.

57. Adams, *Wellborn Science*, 221.

58. A. H. Sturtevant, *A History of Genetics* (New York: Harper and Row, 1965), 4.

59. Heck, "Breeding Back of the Aurochs," 122.

60. Correspondence with Cis van Vuure, Wageningen University, Holland, June 27, 2000.

Chapter 12. A Shot in the Arm

1. Edward Jenner, *Letters of Edward Jenner*, Genevieve Miller, ed. (Baltimore: Johns Hopkins University Press, 1983), 39.

2. Jenner, *Letters*, 4.

3. Jenner, *Letters*, 8.

4. Joel N. Shurkin, *The Invisible Fire: The Story of Mankind's Victory over the Ancient Scourge of Smallpox* (New York: G. P. Putnam's Sons, 1979), 135.

5. Donald Hopkins, *Princes and Peasants* (Chicago: University of Chicago Press, 1983), 79.

6. Hopkins, *Princes and Peasants*, 80.

7. Derrick Baxby, *Jenner's Smallpox Vaccine: The Riddle of Vaccinia Virus and Its Origin* (London: Heinemann Educational Books, 1981), 37.

8. *A Letter to Lord Henry Petty on Coercive Vaccination*, John Stuart, London, 1807. In Jenner, *Letters*, 38.

9. Baxby, *Jenner's Smallpox Vaccine*, vi.

10. Shurkin, *Invisible Fire*, 32.

11. Hopkins, *Princes and Peasants*, 85.

12. Hopkins, *Princes and Peasants*, 88.

13. Baxby, *Jenner's Smallpox Vaccine*, 3.

14. Baxby, *Jenner's Smallpox Vaccine*, 178.

15. Hopkins, *Princes and Peasants*, 83.

16. Baxby, *Jenner's Smallpox Vaccine*, 16.

17. Laurie Garrett, *The Coming Plague: Newly Emerging Diseases in a World Out of Balance* (New York: Farrar, Straus, Giroux, 1994), 381.

18. Roy Porter, *The Greatest Benefit to Mankind: A Medical History of Humanity* (New York: W. W. Norton, 1997), 18.

19. Ann Giudici Fettner, *Viruses: Agents of Change* (New York: McGraw-Hill Publishing Co., 1990), 151.

20. Diamond, *Guns, Germs, and Steel*, 206.

21. Garrett, *Coming Plague*, 447.

22. Garrett, *Coming Plague*, 453.

23. William H. McNeill, *Plagues and Peoples* (New York: History Book Club ed., 1993), 101.

24. McNeill, *Plagues and Peoples*, 247.

25. McNeill, *Plagues and Peoples*, 248.

26. S. Kirnowordoyo, "Zooprophylaxis As a Useful Tool for Control of A. Aconitus Transmitted Malaria in Central Java, Indonesia," *Journal of Communicable Diseases* 18:2 (1986), 90; S. Hewitt, M. Kamal, N. Muhammad, and M. Row-

land, "An Entomological Investigation of the Likely Impact of Cattle Ownership on Malaria in an Afghan Refugee Camp in the North West Frontier Province of Pakistan," *Medical and Veterinary Entomology*, 2 (Apr. 8, 1994), 164.

27. Hopkins, *Princes and Peasants*, 33.

28. Fettner, *Viruses: Agents of Change*, 105.

29. Jenner, *Letters*, 83.

Chapter 13. Mad at Cows

1. Elizabeth Piper, "UK Farmers Say They Have Learned Lessons from BSE," *Reuters America Online*, October 25, 2000.

2. Elizabeth Judge and Valerie Elliott, "Alert on Village's Cluster of CJD Deaths" (London) *Times*, July 14, 2000.

3. Andrew Norfolk, "Life in a 'Wonderful Village' Defies CJD" (London) *Times*, July 14, 2000.

4. Ibid.

5. R. G. Will, A. Alperovitch, S. Poser, M. Pocchiari, A. Hofman, E. Mitrova, R. de Silva, M. D'Alessandro, N. Delasnerie Laupretre, I. Zerr, and C. van Duijn, "Descriptive Epidemiology of Creutzfeldt-Jakob Disease in Six European Countries, 1993–1995," *Annual Neurology* 43:6 (1998), 763.

6. Peta Bee, "How Close Is a Cure for CJD?" (London) *Times*, May 11, 2000.

7. Nigel Hawkes, "Half CJD Victims in Twenties" (London) *Times*, March 29, 2000.

8. "CJD Deaths in Britain Linked to School Meals," *Reuters America Online*, July 16, 2000.

9. Nick Nuttall, "Infected Brain Surgery Tissue Leaves 65 Japanese with CJD" (London) *Times*, April 5, 2000.

10. Emily Green, "A Wonder Drug That Carried the Seeds of Death," *Los Angeles Times*, May 21, 2000.

11. Ibid.

12. Berkow, *Merck Manual*, 922.

13. Elaine Jarvik, "Deer Disease and CJD in Humans," *Deseret News*, January 18, 1999.

14. R. G. Will and R. H. Kimberlin, "Creutzfeldt-Jakob Disease and the Risk from Blood or Blood Products," *Vox Sang* 75:3 (1998), 178.

15. M. P. Busch, S. A. Glynn, and G. B. Schreiber, "Potential Increased Risk of Virus Transmission Due to Exclusion of Older Donors Because of Concern over Creutzfeldt-Jakob Disease: The National Heart, Lung, and Blood Institute Retrovirus Epidemiology Donor Study," *Transfusion* 37:10 (1997), 996.

16. *New York Times*, Mar. 4, 2001, 3-1.

17. L. Tan, M. A. Williams, M. K. Khan, H. C. Champion, and N. H. Nielson, "Risk Transmission of Bovine Spongiform Encephalopathy to Humans in the United States; Report of the Council on Scientific Affairs, American Medical Association," *Journal of the American Medical Association* 281:24 (1999), 2330.

18. Mark M. Robinson, William J. Hadlow, Tami P. Huff, Gerald A. H. Wells, Michael Dawson, Richard F. Marsh, and John R. Gorham, "Experimental Infec-

tion of Mink with Bovine Spongiform Encephalopathy," *Journal of General Virology* 75 (1994), 2151.

19. Jarvik, "Deer Disease and CJD in Humans."

20. Umberto Agrimi, Giuseppe Ru, Franco Cardone, Maurizio Pocchiari, and Maria Caramelli, "Epidemic of Transmissible Spongiform Encephalopathy in Sheep and Goats in Italy," *The Lancet* 353 (Feb. 13, 1999), 561.

21. Green, "A Wonder Drug That Carried the Seeds of Death."

22. "Growth Hormone Blunder May Have Given Britain Mad Cow Disease," *London Observer*, Aug. 8, 1999.

23. Karen Post, Detlev Riesner, Voker Walldorf, and Heinz Mehlhorn, "Fly Larvae and Pupae As Vectors for Scrapie," *The Lancet* 354 (Dec. 4, 1999), 9194.

24. "A Veterinary Scientist Has Proposed a New Theory for the Origin of Mad Cow Disease," *Food Institute Report*, 2000.

25. Chandra Wickramasinghe and FRS Sir Fred Hoyle, "Comets a Possible Source of BSE?" (London) *Times*, Dec. 5, 2000.

26. John R. Fisher, "British Physicians, Medical Science, and the Cattle Plague, 1865–66," *Bulletin of the History of Medicine* 67 (1993), 655.

27. Fisher, "British Physicians," 662.

28. Fisher, "British Physicians," 667.

29. Fisher, "British Physicians," 663.

30. Richard Mandel, "Cow Power," *Engineer's Digest*, 2000.

31. "France Hopes to Use Banned Animal Feed As Fuel," *Reuters America Online*, Dec. 1, 2000.

32. Tina Caparella, "Research Top Priority at Annual Meeting," *Render Magazine*, December 2000, 8.

33. Jerri Bartholomew, personal correspondence with author, March 2001.

Chapter 14. Carnivore Culture

1. Rasmussen, *Agriculture in the United States*, IV, 3259.

2. Eaton, et al., *Paleolithic Prescription*, 5. Jo Robinson, *Why Grassfed Is Best* (Vashon, Wash.: Vashon Island Press, 2000). Artemis P. Simopoulos and Jo Robinson, *The Omega Diet* (New York: HarperCollins, 1999).

3. "Caveman Diet," *Psychology Today* (1997), 18.

4. A. F. Subar, "Dietary Sources of Nutrients Among U.S. Children, 1989–1991," *Pediatrics* 102:4 (1998), 913.

5. John R. Campbell and John F. Lasley, *The Science of Animals That Serve Mankind* (New York: McGraw-Hill Book Co., 1969), 27.

6. Reay Tannahill, *Food in History* (New York: Crown Publishers, 1988), 149.

7. Fernand Braudel, *The Structures of Everyday Life: The Limits of the Possible* (New York: Harper and Row, 1981), 190.

8. Braudel, *Structures of Everyday Life*, 192.

9. Kenneth J. Carpenter, *Protein and Energy: A Study of Changing Ideas in Nutrition* (New York: Cambridge University Press, 1994), 60.

10. Carpenter, *Protein and Energy*, 74.

11. Florence Nightengale, *Notes on Nursing: What It Is, and What It Is Not* (New York: Dover Publications, 1969), 69.

12. Nightengale, *Notes on Nursing*, 79.

13. Carpenter, *Protein and Energy*, 81.

14. Carpenter, *Protein and Energy*, 83.

15. J. H. Kellogg, *Colon Hygiene* (Battle Creek, Mich.: Good Health Publishing, 1916), 9.

16. Kellogg, *Colon Hygiene*, 10.

17. Kellogg, *Colon Hygiene*, 62.

18. Kellogg, *Colon Hygiene*, 160.

19. Kellogg, *Colon Hygiene*, 87.

20. F. S. Nitti, "The Food and Labor Power of Nations," *Economics Journal* 6 (1896), quoted in Carpenter, *Protein and Energy*, 97.

21. Carpenter, *Protein and Energy*, 114.

22. Ibid.

23. Carpenter, *Protein and Energy*, 109.

24. Carpenter, *Protein and Energy*, 105.

25. Carpenter, *Protein and Energy*, 118.

26. Ibid.

27. Francis Moore Lappé, *Diet for a Small Planet* (New York: Ballantine Books, 1971), 40.

28. Carpenter, *Protein and Energy*, 179.

29. Carpenter, *Protein and Energy*, 203.

30. Carpenter, *Protein and Energy*, 232.

31. Carpenter, *Protein and Energy*, 234.

32. Jeremy Rifkin, *Beyond Beef: The Rise and Fall of the Cattle Culture* (New York: Penguin Books, 1992), back jacket copy.

33. Mark Friedberger, "Cattlemen, Consumers, and Beef," *Environmental History Review* 18:3 (1994), 39.

34. Ibid.

35. Friedberger, "Cattlemen, Consumers, and Beef," 48.

36. Calvin Tomkins, "Table Talk: Two Classic Chefs Want to Banish America's Fear of Fat," *New Yorker*, Nov. 8, 1999, 32.

Chapter 15. The Milk Cow

1. William Dempster Hoard, website for *Hoard's Dairyman*, Internet at: www.hoards.com/history/wdh.html.

2. Toussaint-Samat, *History of Food*, 113.

3. Toussaint-Samat, *History of Food*, 114.

4. Lena Sommestad, "Gendering Work, Interpreting Gender: The Masculinization of Dairy Work in Sweden, 1850–1950," *History Workshop Journal*, 37 (1994), 60.

5. Toussaint-Samat, *History of Food*, 116.

6. Toussaint-Samat, *History of Food*, 117.

7. Ibid.

8. Campbell and Lasley, *Science of Animals That Serve Mankind*, 264.
9. Beatrice Fontanel and Claire d'Harcourt, *Babies: History, Art, and Folklore* (New York: Harry N. Abrams, 1997), 92.
10. Fontanel and d'Harcourt, *Babies*, 107.
11. Fontanel and d'Harcourt, *Babies*, 113.
12. Fontanel and d'Harcourt, *Babies*, 112.
13. Fontanel and d'Harcourt, *Babies*, 119.
14. Fontanel and d'Harcourt, *Babies*, 121.
15. Fontanel and d'Harcourt, *Babies*, 122.
16. Olwen Hufton, *The Prospect Before Her: A History of Women in Western Europe, 1500–1800* (New York: Alfred A. Knopf, 1996), 77.
17. Rasmussen, *Agriculture in the United States*, 1048.
18. Narcissa Whitman, *My Journal*, ed. Larry Dodd (Fairfield, Wash.: Ye Galleon Press, 1994), 17.
19. Whitman, *My Journal*, 20.
20. Whitman, *My Journal*, 23.
21. Clifford Merrill Drury, *The Diaries and Letters of Henry H. Spalding and Asa Bowen Smith Relating to the Nez Perce Mission, 1838–1842* (Glendale, Calif.: Arthur H. Clark Co., 1958), 234.
22. Whitman, *My Journal*, 52.
23. Clifford Merrill Drury, *First White Women over the Rockies*, vol. II (Glendale, Calif.: Arthur H. Clark, Co.), 214.
24. Drury, *First White Women over the Rockies*, 216.
25. Cornelius J. Brosnan, *Jason Lee: Prophet of the New Oregon* (New York: Macmillan, 1932), 87.
26. Kenneth L. Holmes, *Ewing Young: Master Trapper* (Portland, Ore.: Binfords and Mort, 1967), 129.
27. David Lavender, *Land of Giants: The Drive to the Pacific Northwest, 1750–1950* (Lincoln: University of Nebraska Press, 1958), 188.
28. Lavender, *Land of Giants*, 375.
29. Rasmussen, *Agriculture in the United States*, 1680.
30. Rasmussen, *Agriculture in the United States*, 1681.
31. Rasmussen, *Agriculture in the United States*, 1682.
32. Rasmussen, *Agriculture in the United States*, 1683.
33. Rasmussen, *Agriculture in the United States*, 1685.
34. Rasmussen, *Agriculture in the United States*, 1686.
35. Jay Stuller, "In the Land of Milk Money," *Smithsonian*, March 1999, 2.
36. Ibid.
37. Ibid.
38. Ibid.
39. Donna Gorski, "Heart-Healthy Milk," *Dairy Foods* (1998), 1.
40. Gene Johnston, "It's a Mini Dairy: New Milk Processing System Lets You Add Value on the Farm," *Successful Farming*, January 1999, 1.

CHAPTER 16. RUMINATIONS: CARE AND FEEDING

1. Campbell and Lasley, *Science of Animals That Serve Mankind*, 645.

2. Campbell and Lasley, *Science of Animals That Serve Mankind*, 625.

3. Campbell and Lasley, *Science of Animals That Serve Mankind*, 47.

4. Internet at: www.ansi.okstate.edu/BREEDS/.

5. Tilden Wayne Perry and Michael J. Cecava, *Beef Cattle Feeding and Nutrition* (New York: Academic Press, 1995), 156.

6. Perry and Cecava, *Beef Cattle Feeding*, 156–164.

7. Eric R. Haapapuro, Neal D. Barnard, and Michele Simon, "Animal Waste Used As Livestock Feed: Dangers to Human Health," *Preventive Medicine* 26:5 (1997), 598.

8. Haapapuro, et al., "Animal Waste," 599.

9. Lynn B. Jacobs, *Waste of the West: Public Lands Ranching* (Tucson, Ariz.: L. Jacobs, 1991), 3.

10. Jacobs, *Waste of the West*, 9, 10.

11. Jacobs, *Waste of the West*, 11.

12. Ed Marston, "Experts Line Up on All Sides of the Tree-Grass Debate," *High Country News* 28:7 (1996).

13. Wendee Holtcamp, "Cows in the Forest," *American Forests* 104:2 (1998).

14. Maryanne Vollers, "Buffalo Soldiers," *Mother Jones* (Nov./Dec. 1999), 75.

15. Vollers, "Buffalo Soldiers," 78.

16. Jomo Kenyatta, "An African Point of View," *New Statesman* (Nov. 29, 1999), Letters to the Editor.

17. J. S. Perkins, "Botswana: Fencing Out the Equity Issue: Cattleposts and Cattle Ranching in the Kalahari Desert," *Journal of Arid Environments* 33:4 (1996), 513.

18. J. S. Perkins, "Botswana," 517.

19. "60 Die in Cattle Herders' Clash," *New York Times*, May 3, 2000.

20. Jacobs, *Waste of the West*, 564.

21. Will Baker, *Tony and the Cows: A True Story from the Range Wars* (Lewiston, Idaho: Confluence Press, 2000), 41.

22. Baker, *Tony and the Cows*, 68.

23. Baker, *Tony and the Cows*, 107.

24. Baker, *Tony and the Cows*, 108.

25. Baker, *Tony and the Cows*, 119.

26. Barney Nelson, *The Wild and the Domestic: Animal Representation, Ecocriticism, and Western American Literature* (Reno: University of Nevada Press, 2000), xi.

27. Nelson, *The Wild and the Domestic*, 1.

28. Nelson, *The Wild and the Domestic*, 7.

29. Nelson, *The Wild and the Domestic*, 15.

30. Budiansky, *Covenant of the Wild*, 167.

Index

A NOTE ON THE AUTHOR

Laurie Winn Carlson's interests have frequently centered on the American West. Born in Sonora, California, she studied at the University of Idaho, Arizona State University, and Eastern Washington University. Her other books include *A Fever in Salem*, a widely praised reinterpretation of the New England witch trials; *On Sidesaddles to Heaven: The Women of the Rocky Mountain Mission*; and the award-winning children's book *Boss of the Plains: The Hat That Won the West*. She is married with two sons and lives in Cheney, Washington.